AF389007

Maternal DHA Impact on Child Neurodevelopment

Maternal DHA Impact on Child Neurodevelopment

Editor

Asim K. Duttaroy

MDPI • Basel • Beijing • Wuhan • Barcelona • Belgrade • Manchester • Tokyo • Cluj • Tianjin

Editor
Asim K. Duttaroy
University of Oslo
Norway

Editorial Office
MDPI
St. Alban-Anlage 66
4052 Basel, Switzerland

This is a reprint of articles from the Special Issue published online in the open access journal *Nutrients* (ISSN 2072-6643) (available at: www.mdpi.com/journal/nutrients/special_issues/DHA_cognition).

For citation purposes, cite each article independently as indicated on the article page online and as indicated below:

LastName, A.A.; LastName, B.B.; LastName, C.C. Article Title. *Journal Name* **Year**, *Volume Number*, Page Range.

ISBN 978-3-0365-1616-5 (Hbk)
ISBN 978-3-0365-1615-8 (PDF)

Contents

About the Editor

Asim K. Duttaroy

Dr. Asim K. Duttaroy is a Professor and Group Leader at the Clinical Nutrition at the Faculty of Medicine, University of Oslo, Norway. His main research area is feto-placental growth and development, and the cardioprotective function of naturally derived molecules. His scientific work has been published in over 270 original contributions, reviews, books, book chapters, editorials, and 9 books. He has successfully supervised 35 PhD/MD students so far. He has examined more than 270 PhD theses submitted from different countries, including India, the USA, the UK, South Africa, and Australia. He has several international patents that led to the establishment of two multinational companies, Provexis Ltd., UK, and IDIA AS, Norway. His patented work on the anti-platelet regime (Fruitflow®) isolated from tomatoes, an alternative to aspirin, was the EU's first approved EFSA product. He received numerous international awards for his contribution to clinical research.

Preface to "Maternal DHA Impact on Child Neurodevelopment"

This preprint book focuses on maternal docosahexaenoic acid, 22:6n-3 (DHA), and arachidonic acid, 20:4n-6 (ARA), on children's neurodevelopment.

I express my gratitude to all authors for their excellent contributions from mechanical aspects to interventional studies.

I am also grateful to Ms. Stella Duo, the managing editor, the journal *Nutrients*, MDPI, and all of the people involved in making this a successful Special Issue on maternal DHA impact on child neurodevelopment.

I hope the readers find the papers in this preprint book very helpful as they contain new information.

Asim K. Duttaroy
Editor

 *nutrients*

Review

Maternal Docosahexaenoic Acid Status during Pregnancy and Its Impact on Infant Neurodevelopment

Sanjay Basak, Rahul Mallick and Asim K. Duttaroy

1   Molecular Biology Division, National Institute Nutrition, Indian Council of Medical Research, Hyderabad 500 007, India; sba_bioc@yahoo.com
2   Department of Biotechnology and Molecular Medicine, A.I. Virtanen Institute for Molecular Sciences, University of Eastern Finland, 70211 Kuopio, Finland; rahul.mallick@uef.fi
3   Department of Nutrition, IMB, Faculty of Medicine, University of Oslo, 0317 Oslo, Norway
*   Correspondence: a.k.duttaroy@medisin.uio.no; Tel.: +47-22-82-15-47

Received: 19 October 2020; Accepted: 23 November 2020; Published: 25 November 2020

Abstract: Dietary components are essential for the structural and functional development of the brain. Among these, docosahexaenoic acid, 22:6n-3 (DHA), is critically necessary for the structure and development of the growing fetal brain *in utero*. DHA is the major n-3 long-chain polyunsaturated fatty acid in brain gray matter representing about 15% of all fatty acids in the human frontal cortex. DHA affects neurogenesis, neurotransmitter, synaptic plasticity and transmission, and signal transduction in the brain. Data from human and animal studies suggest that adequate levels of DHA in neural membranes are required for maturation of cortical astrocyte, neurovascular coupling, and glucose uptake and metabolism. Besides, some metabolites of DHA protect from oxidative tissue injury and stress in the brain. A low DHA level in the brain results in behavioral changes and is associated with learning difficulties and dementia. In humans, the third trimester-placental supply of maternal DHA to the growing fetus is critically important as the growing brain obligatory requires DHA during this window period. Besides, DHA is also involved in the early placentation process, essential for placental development. This underscores the importance of maternal intake of DHA for the structural and functional development of the brain. This review describes DHA's multiple roles during gestation, lactation, and the consequences of its lower intake during pregnancy and postnatally on the 2019 brain development and function.

Keywords: DHA; brain; MFSD2a; SPM; fetus; placenta; infant; neurogenesis; pregnancy; pre-term

---

1. Introduction

Docosahexaenoic acid, 22:6n-3(DHA), an n-3 (omega-3) long-chain polyunsaturated fatty acid (LCPUFA), plays various essential roles in human health [1–4]. DHA takes part in several biological processes, such as angiogenesis, immune modulation, inflammatory response, signal transduction, apoptosis, cell proliferation, and a host of the membrane, cellular and molecular functions that affect health and diseases [3,5,6]. The DHA metabolites also mediate their roles in cellular signaling processes [7]. Inadequate dietary consumption of DHA and eicosapentaenoic acid, 22:5n-3 (EPA) impairs the optimal growth of the feto-placental unit and imposes plausible risks of cognitive decline, inflammatory disease, cardiovascular disease, inflammatory disorders, behavioral changes, and mental stress in later life [5,8,9]. While the metabolites of EPA and DHA, such as eicosanoids and docosanoids, are involved in cell signaling, DHA is primarily used for membrane structure and function [3,4]. Modern refined diets are mostly deficient in n-3 PUFAs, leading to sub-optimal organ function that may predispose individuals to an increased risk of diseases [9–11].

The recommended intake of DHA and the EPA is based on the quantum of data supporting beneficial outcomes in protecting cognitive development and cardiovascular diseases. Experts are agreed upon to enhance n-3 LCPUFA intake via increased seafood or supplementary approaches. An n-3 LCPUFA intake of no less than 250–500 mg/d should be made available to maintain healthy adults' physiological needs in their daily routine. Even though DHA consumption during pregnancy is an essential consideration as it is required for fetal brain development and growth, a limited number of countries have adopted and implemented appropriate guidelines for the n-3 LCPUFA consumption during pregnancy. DHA's requirement during placental growth and early development has been highlighted recently. Thus, low maternal DHA status may lead to the placenta's functional inadequacy and accordingly alter fetoplacental growth and development. Therefore, DHA supplementation well before gestation can be considered to prevent feto-placental associated developmental disorders.

Several comprehensive articles have focused on DHA and its impacts on human health [3,4,12–14]. However, this review has primarily concentrated on maternal DHA status during pregnancy on neurocognitive development in infants.

2. DHA and Its Metabolites: Effects on the Structure and Function of the Human Brain

N-3 PUFAs have essential functions on human health, and various studies have explained the molecular mechanisms underlying the effects of DHA in various tissues, including the brain. DHA is the predominant n-3 LCPUFA within the brain. Several neurophysiological functions are attributed to DHA, including the cell-survival, neuroinflammation, neurogenesis cellular signaling, and its protective function in maintaining blood-brain barrier integrity. Because of these vital roles of DHA in the brain, any alteration of DHA metabolism in the brain, as a cause or consequence, affects several neurological and psychiatric conditions.

DHA derivatives include the families of specialized pro-resolving mediators (SPMs), such as lipoxins, resolvins (resolvin D (RvD), and resolvin E (RvE)), protectins, and maresins, are known to have diverse biological activities [15,16]. SPMs actively thwarted the inflammatory response by stimulating specific G-protein-coupled receptors (GPR) expressed on immune cells that propagate both the anti-inflammatory and pro-resolving processes [15,16]. The anti-inflammatory activities of SPMs are mediated by inflammatory scavenging interleukins (IL) such as IL-10, IL-1 decoy receptors, and IL-1 receptor antagonists and anti-inflammatory cytokines [15,16]. The SPMs activate their actions by several mechanisms that include the anti-inflammatory resolution, blocking the intracellular pathways leading to inflammation, downregulation of pro-inflammatory cytokines, and clearance of inflammatory cell debris macrophages, as well as stabilization of immune cells counts to basal levels [15,16]. The importance of SPMs in the inflammatory resolution is widely reported in chronic pathologies, including brain inflammation when their production remained insufficient, and the administration of SPMs exogenously reduces the inflammatory process and protects inflamed tissue(s) [15,17]. Dietary supplementation with metabolic precursors of SPMs can increase their availability and resolve the inflammatory process following neurological injury.

In the brain, DHA and arachidonic acid, 20:4n-6 (ARA) are incorporated into cell membrane phospholipids and thus influence their metabolism. ARA represents approximately 20% of the total amount of fatty acids in neurons. ARA, primarily esterified in phospholipids of the membranes, is released by the phospholipases A_2. Several enzymes convert free ARA into several eicosanoids, and these metabolites may be involved in neuroinflammation [18]. Unesterified ARA can also directly modify synaptic functions [19]. Therefore, the level of intracellular free ARA plays a critical role in both neuroinflammation and synaptic functions. Inflammation may prolong due to the presence of ARA-derived eicosanoids, such as prostaglandin E_2 and leukotriene B_4, whereas n-3-derived eicosanoids decrease the levels of these inflammatory compounds and their activity. Thus, supplementation of n-3 fatty acids can restore or reverse the balance of lipid mediator precursors. Metabolic precursors of the inflammation-resolving SPMs, including DHA itself, having vital inflammation resolution effects in brain injury [20]. Intravenously administered low-dose DHA in rats showed significant tissue-sparing effects in the peri-infarct region of the middle cerebral artery occlusion compared to those who received high or medium DHA doses [21]. DHA-treated rats improved significantly in neurological performance up to 7 days following middle cerebral artery occlusion. Following experimental traumatic brain injury in rats, a DHA-enriched diet for 12 days preserved brain-derived neurotrophic factor (BDNF) concomitant improved learning capability [20]. Although DHA's mechanism in rescuing neurological injury is not known, its potential therapeutic warrants further investigations.

Resolvins, protectins, and maresins have SPM functions. SPMs are produced from DHA either spontaneously or by aspirin, involving enzymes, such as lipoxygenase (LOX) and acetylated cyclooxygenase-2 (COX-2) enzymatic pathway. Two distinct classes of the resolvins (Rvs) are synthesized from DHA by different biosynthetic paths, denoted as 17S- and 17RD- series resolvins during inflammation of the resolution [22,23]. The strong anti-inflammatory and pro-resolving activities of Rvs were demonstrated in animal models [7]. Rv of E and D series are endogenously produced from EPA and DHA, respectively. Both RvD1 and RvE1 may potentially lower inflammation. Conversion of DHA to 17(S)-hydroxy-containing RvD1–D4 and conjugation to triene-containing docosanoid structures are carried out by LOX [24]. Rvs can reduce the augmented pain by regulating the mediators of inflammation, ion channels with transient receptor potential, and transmission via the spinal cord. RvD_1 (17(S)-trihydroxy-docosahexaenoic acid) is structurally different from aspirin-triggered RvD_1. COX-2 promotes the synthesis of 17(R)-trihydroxy-docosahexaenoic acid, followed by the LOX on 13-hydroxyDHA [25]. The transcription of IL-1β is prevented by RvD1 that reduces the infiltration of neutrophils into the brain [23]. Since all Rvs are potent anti-inflammatory agents, dietary supplementation of n-3LCPUFAs is beneficial in the inflammatory situation.

The anti-inflammatory protectin D_1 (PD1) is also synthesized endogenously from DHA in neuronal tissues and, therefore, termed neuroprotectin (NPD_1). The internally formed NPD_1,17(S)-trihydroxy DHA varies from aspirin-triggered NPD_1; 17(R)-trihydroxy DHA structurally. Protectins are anti-microbial agents. PD_1 was shown to prevent infiltration PMN both in vivo and in vitro. NPD_1 effectively prevents damage to several mice tissues, such as the retina, brain, liver, kidney, and fibrosis [26]. The PD1/NPD1 isomer of PDX can prevent both inflammation and atherogenesis [27,28]. Besides its insulin-sensitizing and glucose regulatory actions, PDX can inhibit inflammation via IL-6 synthesis [29,30]. Maresins, MaR (macrophage mediators in resolving inflammation) are endogenously synthesized from DHA by macrophages through 12-LOX, followed by epoxy hydrolase activity. Maresins have both anti-inflammatory and pro-resolving actions [31]. The two structural variants of MaR are MaR 1 and MaR 2. MaR 1 (7R,14S-dihydroxy-docosa-4Z,8E,10E,12Z,16Z,19Z-hexaenoic acid) is a family of structurally distinct autacoids [32]. Studies on MaR 1 has potent anti-inflammatory activity even at the nano molar range [33].

Resolvins and protectins are produced from EPA and DHA. Both resolvins and protectins have an anti-inflammatory function. While 18R E-series resolvins (RvE1 and RvE2) are generated from EPA, DHA-derived 17S D-series resolvins (RvD1 to D6), PD1, and MaRs are synthesized during the inflammatory resolution. These inflammation preventive mediators are critically involved in resolving inflammation for the healing process. These EPA- and DHA-derived anti-inflammatory compounds inhibit trans-endothelial migration of PMN, suppress dendritic cell migration, reduce leukocyte infiltration [34]. PD1 blocks the synthesis of T cells-derived tumor necrosis factor alpha TNF-α and interferon gamma IFN-γ and induces apoptosis of these cells, implying its role in the down-regulation of Th1-mediated responses. LOX-dependent Th2-skewed human peripheral blood mononuclear cells-derived PD1 may favor the Th2 phenotype [35]. Several studies suggest that both n-3LCPUFAs, DHA, EPA, and their inflammation preventive mediators, such as MaRs, PDs, and Rvs, play vital role in mitigating the inflammatory process and support wound healing. Therefore, these bioactive lipids could be developed to manage and prevent various inflammatory diseases [34].

2.1. DHA Accretion, Supplementation, and Fetal Brain Development

DHA is vital for the development of a healthy brain [12,36,37]. Quantitatively in the brain, DHA is considered as the vital fatty acid [38]. In the human brain, the DHA level is 250–300 times higher than EPA. DHA is predominantly found in the phospholipid fraction of the brain grey matter [39]. The phospholipid distribution of n-3 LCPUFA in the brain showed a striking difference between EPA and DHA. DHA is primarily present in phosphatidylethanolamine and phosphatidylserine fractions, whereas EPA is mostly distributed in the phosphatidylinositol fraction of the membrane phospholipids. DHA is the most abundant n-3 fatty acids in the entire nervous system. DHA is critically required in the neuron regeneration and formation of synapse during the fetus's development and the first two years following birth [40].

Accumulation of DHA in the fetal brain occurs continuously throughout the gestation but is most active during week 29 to week 40. Supplementation of DHA (200 mg DHA/day) during the third trimester of pregnancy prevented the decrease of maternal DHA status [41]. Due to continuum fetal DHA accretion, the nutritional status of maternal DHA during pre-conception, pregnancy, and lactation demands DHA requirements for the brain and retinal development of babies [42,43]. Neonates with higher DHA concentrations in umbilical plasma phospholipids showed a longer gestational length than neonates with a lower concentration of DHA [44]. The pregnant women, supplemented with 600 mg DHA/day before 20 weeks of gestation until delivery, significantly reduced the preterm delivery and low-birth-weight babies [45]. DHA supplementation improves the DHA status both in the mother and her child because of its efficient transfer through the placenta [46], and breast milk [47,48]. DHA supplementation was beneficial for mothers having low consumption of seafoods [49]. DHA supplementation during pregnancy increases the placental transport of n-3 LCPUFAs via increased expression of fatty acid-binding/ transport proteins [50,51]. Consuming DHA (2.2 g DHA/day) and EPA (1.1 g EPA/day), from the 20th week of pregnancy until the partum showed an improved visual and coordination capacity of the children [52]. Similar beneficial effects were obtained after the supplementation of mothers with 500 mg DHA/day during pregnancy, which correlated the high blood DHA levels with the improved cognitive development of 5.5-year-old children [53]. Similarly, daily supplements of fish oil (500 mg DHA + 150 mg EPA), along with 5-methyltetrahydrofolate (400 µg/day), showed improved cognitive development until 6.5 years of age [54]. Higher DHA levels in plasma and breast milk positively correlate the brain's growth, development, and visual acuity in the neonate [55,56]. These studies demonstrated that DHA supplementing mothers during pregnancy and lactation or baby food fortified with DHA enhanced DHA levels in the infant tissues with improved neurological and visual development [57]. Conversely, a diet low in n-3 LCPUFA during pregnancy and/or lactation may negatively impact the child's visual and neurological development [58,59].

Infants fed with breast milk (<0.17% of total fatty acid in milk in contrast to optimal 0.3–0.4% DHA) showed a lower amount of DHA in erythrocytes, reduced visual acuity, and delayed language development as compared to infant fed by breast milk containing 0.36% DHA [43,60]. Women who received DHA (600 mg/day) supplementation during pregnancy from <20 weeks until delivery showed a substantial increase in visual acuity, particularly in male newborns. DHA supplementation may be the best predictor for nervous system development [61]. A direct relationship between higher DHA levels in erythrocytes (in mother and children) and the optimal visual and neuronal development of children was reported [62–65]. Perinatal DHA supplementation was also reported to reduce the risk of lower intelligence quotient (IQ) scores in children from very low-income families [66,67]. Newborn, when supplemented with DHA enriched formulas, improved cortical maturation and visual function [68]. Six-month-old babies who did not receive maternal milk and fed a formula containing egg yolk enriched with DHA (115 mg DHA/100 g food) showed a significant increase in erythrocyte DHA phospholipids and a better visual development measured at 1 year [69].

Supplementation of a minimum of 0.35% DHA formulated foods favors improved brain development until 4 months after delivery, as evaluated by the term's mental development index [70]. In addition to DHA's dietary intake, ARA is obligatory for optimal brain and eye development in infants [71]. The child who did not receive maternal milk must be fed with a formula containing DHA and ARA for optimum brain growth and development until 39 months after birth [72] and even 4 years after birth [71]. A study in newborns ($n = 343$) of 1–9-day-olds fed with formula with varying levels of DHA ((control (DHA 0%); 0.32%; 0.64%; 0.96%, with ARA 0.64% in all formulas) until the 12th month showed that babies fed with formulas containing 0.32% DHA showed a significant improvement in cognitive development as compared to control group [73]. An improved problem-solving skills and memory were noticed when human milk supplemented in DHA (32 mg/day) and ARA (31 mg/day) were fed to the preterm infant (birth weight < 1500 g) until 9 weeks after birth [74]. DHA (0.05 g/100 g) and ARA (0.1 g/100 g) fortified feeding to the preterm infant (birth weight > 2000 g) performed a higher index in mental development scores [75]. The term newborn ($n = 420$) supplemented with n-3 LCPUFA (60 mg of EPA and 250 mg of DHA daily) for six months following birth showed a significant increase of DHA deposition in the erythrocyte phospholipids and language and communication skills development [76]. A follow-up study with a formula (0.5% DHA) feeding preterm children ($n = 107$) from their birth until nine months showed a significant improvement in the verbal and total intellectual coefficient, language capacity, and memory in girl children as compared to control (0.0% DHA) [77]. A comparative study among the infants fed on breast milk, a preterm formula supplemented with LCPUFAs, or a traditional preterm formula without LCPUFAs, showed that preterm formulas with LCPUFAs improve the visual understanding and development of infants like those fed with maternal milk [78]. A large study involving 28 countries demonstrated that better mathematics test scores in children from low-income families are proportionate to the DHA levels in breastfed children's performance was found superior to non-breastfed children from high-income families and/or increased spending on education [79].

DHA level in the blood is positively correlated to cognitive abilities and inversely associated with cognitive decline. Activation of membrane GPR40 receptor by LCPUFA is responsible for driving neurogenesis by modulating synaptic plasticity in the adult brain. Dysfunction of GPR40 receptors can produce lipotoxicity in brain endothelial cells and neurons [80]. LCPUFA mediate GPR40 signaling modulates neurogenesis's functional aspects, anti-nociceptive effects, anti-apoptotic effect, and Ca2+ homeostasis in Alzheimer's disease (AD), and the nigrostriatal pathways. GPR40 is chosen as a potential drug target in treating several neuropathological disorders, including AD. During brain development, the DHA availability is also influenced by genetic polymorphisms of enzymes responsible for the endogenous conversion to LCPUFAs from their precursors.

Linoleic acid, 18:2n-6 (LA), and alpha-linolenic acid,18:3n-3 (ALA), are the two essential fatty acids. Imbalanced dietary intake of n-3 and n-6 fatty acids may lead to several neurological conditions. Brain concentrations of n-3 and n-6 PUFA derivatives are modulated by their dietary level and play a central role in regulating mood and cognition [81]. Endogenous abilities in the conversion of n-3 LCPUFAs from its precursor is affected by altered gene polymorphisms. The presence of gene-specific polymorphisms encodeΔ-5 and Δ-6 desaturases, enzymes involved in the synthesis of n-3 LCPUFAs from the precursor ALA, altered these fatty acid levels in particular DHA [82]. The f gene polymorphism (rs 174575) of Δ-6 desaturase enzyme in children allows higher DHA levels and better IQ test scores [83]. This indicates the importance of gene variations in the metabolism of n-3 LCPUFA and a consequential beneficial effect on brain development. Nine-month-old children supplemented with high DHA content showed significant cognitive performance and better arterial pressure at the end of infancy. N-3 LCPUFA supplementation, especially DHA, may favor optimum brain development that also has a protective effect against adult life's cognitive and cardiovascular diseases [84]. Supplementation of DHA (400 mg/ daily) for 4 months significantly raised blood DHA levels, positively correlated with an increment of punctuation score obtained for vocabulary and comprehension tests in 4-year-old children [85]. Supplementation of children ($n = 409$, 3 to 13-year-old) in Australia, with 750 mg DHA and 60 mg EPA per day for 20 weeks, demonstrated a significant increment in children's academic performance between 7 and 12 years old [86]. The supplementation of fish oil rich in DHA to 7–12-year-old ADHD children in Australia showed improved word reading, spelling capacity, and parents' conduct. These changes were positively related to an increase in the DHA level of erythrocyte phospholipids [87].

Despite this evidence, absolute compliance with DHA supplementation of baby food has not been established yet. The Scientific Opinion of the European Food Safety Authority (EFSA) Panel (2014) expressed that " . . . *there is no convincing evidence that the addition of DHA to infant and follow-on formulae has benefits beyond infancy on any functional outcomes*". The panel's proposal about DHA supplementation of infant formula is based on DHA's structural and functional roles in nervous tissue and retina and its exclusive presence in the brain and retina's normal development.

2.2. Maternal DHA and Its Effects on the Placental Structure and Functional Development

Inappropriate structure and functional property of the placenta can impair the adequate exchange of nutrient and waste products between materno-fetal compartments during gestation [88]. The placenta's transport efficiency critically determines the fetus's optimal development and the well-being of the mother. Majorities of complicated pregnancies in humans are linked with the altered structure and function of the placenta. Inadequate supply of maternal blood to the fetus via the placenta is correlated adversity with risks of preeclampsia (PE), and intrauterine growth restriction (IUGR), preterm delivery (PT), etc. Shallow invasion of uterine spiral arteries by extravillous trophoblasts (EVTs) is a characteristics phenotype in PE, significant complications of the human pregnancy [89]. The EVTs are engaged in remodeling the uterine spiral artery to improve blood flow from the maternal circulation to the fetoplacental unit. Typically, humans' trophoblast invasion is confined to the endometrium and the first third of the myometrium and restricted within the first trimester of pregnancy. The pathogenesis of PE is primarily caused due to inadequate angiogenesis and placental dysfunction, leading to adversity in pregnancy outcomes for both mother and babies. Angiogenesis forms a new blood vessel from pre-existing ones, critically important for placental development and vasculature [90]. Several growth factors are known to influence the angiogenesis processes that include vascular endothelial growth factor (VEGF), platelet-derived growth factor (PDGF), angiopoietin-like protein 4 (ANGPTL4), platelet-activating factor (PAF), matrix metalloproteinases (MMP), etc. [91].

Recent data highlights about DHA are that in addition to its essential requirement in the third trimester of pregnancy, it is also involved in the first trimester's placentation processes. DHA stimulated tube-like formation in vitro to mimic angiogenesis during the early trimester of pregnancy in first-trimester human placental trophoblast cells, HTR8/SVneo cells [92,93]. Among the long-chain fatty acids tested in vitro, DHA was found the most potent stimulator of the in vitro angiogenesis in placental cells [92,94]. Moreover, DHA stimulated tube formation was partly mediated by increased expression and secretion of VEGFA synthesis, whereas oleic acid OA, ARA, EPA mediated their effects via increased synthesis of ANGPTL4 in these cells. The finding of n-3LCPUFAs on VEGFA synthesis and angiogenesis in placental cells opposed their other cell types' actions [95,96]. The induction in the expression of VEGFA by DHA was found unique in many ways as its expression was induced by several growth factors and cytokines, including EGF, PDGF, TGFβ1, TNFα, and IL-1β, but not by any fatty acids. The mechanism of DHA's action is still to be defined. However, DHA metabolites are unlikely to be involved as DHA induced VEGFA expression was not accompanied by COX-2 expression, a predominant mediator of DHA metabolites. Since peroxisome proliferator-activated receptor (PPAR,) agonist, and ligands failed to stimulate VEGFA expression, the involvement of PPARγ in DHA-stimulated VEGFA expression was unlikely. Both DHA and the EPA showed anti-angiogenic effects in cancer cells by inhibiting angiogenic mediators' production, VEGF, nitric oxide (NO,) PDGF, COX-2, PGE2 [95,97]. EPA enhanced VEGF synthesis by activating GPR120 and PPARγ pathway in 3T3-L1 cells [98]. Although DHA metabolite, such as 4-hydroxy-docosahexaenoic acid (4-HDHA), inhibited PPARγ-mediated angiogenesis in endothelial cells [99], however, DHA metabolites are not involved in the tube formation and/ VEGF synthesis of the placental trophoblast cells as fatty acid receptor genes fatty acid-binding protein-4 (FABP4, GPR120, GPR40) and lipid metabolic genes (COX-2 and CAV-1) is not upregulated by DHA [94]. The precise mechanism of enhanced tube formation by

EPA and DHA in placental trophoblast cells still to be ascertained. These LCPUFAs increased COX-2 messenger RNA mRNA expression, which may play a role in angiogenesis [94]. The mechanism that distinguishes DHA from other LCPUFA may be due to structural differences of the fatty acids. Further data postulates that DHA stimulates the expression of FABP4, an upstream target of VEGF in other cell systems. However, detailed work is essential to underscore DHA's differential effects over other fatty acids in endothelial, tumor, and placental trophoblast cells.

DHA delivery is essential for fetal brain and eye development during the third trimester of pregnancy [93,100]. Due to the limited synthesis and endogenous DHA conversion, the growing fetus relies mostly on the maternal DHA's placental supply [100]. Maternal plasma free fatty acids are taken up by the placental trophoblast via several membranes spanning proteins, such as fatty acid transport proteins (FAT/ cluster of differentiation 36 CD36, FATPs, and plasma membrane fatty acid-binding protein FABPpm) and intracellular fatty acid-binding proteins (FABPs) [101]. The mechanism of DHA uptake and transport in the placenta is not fully understood yet. However, a preferential uptake of DHA transport over other fatty acids via placental plasma membrane fatty acid-binding protein (p-FABPpm) was demonstrated [51,102,103]. Recently, a transporter major facilitator superfamily domain-containing 2A (MFSD2a) is present in the human placenta. MFSD2a) is located in the blood-brain barrier vessel and is involved in transporting the lysophosphatidylcholine form of DHA from the circulation into the brain. The role of MFSD2a in the placenta is not clear at this moment. Expression of the MFSD2a gene during pregnancy is used as a biomarker in predicting fetal neurodevelopment [104]. Maternal MFSD2a levels in blood were positively associated with the Z-score of head circumference at multiple time points at the neonate's first year. Placental MFSD2a transporter expression decreased and correlated to decreased DHA in cord blood of women with gestational diabetes, indicating its role in contributing to materno-fetal DHA transport [51,105,106]. Not much is known about the cellular localization of MFSD2a in the human placenta yet, although its expression is altered in the placenta of gestational diabetes placenta and PE. It is still not known whether the expression of MFSD2a in normal pregnancy may be used as an indicator of DHA deficiency in the fetus's brain.

Several researchers reported an increased incidence of preeclampsia in women with low n-3 fatty acids content in their red blood cells [107]. Increased n-3 LCPUFAs intake during pregnancy was beneficial for overall fetal growth and lowered the risk of early delivery and preeclampsia incidence [2,108]. In preterm birth, a positive association between placental DHA contents with placental weight was reported [109]. Maternal DHA content is, therefore, not only crucial for fetal growth and development but may also contribute a role in determining the placental size by stimulating first trimester trophoblasts mediated early placentation processes. International scientific experts recommend maternal consumption of pre-formed DHA for the prevention of premature birth [110], based on prevailing data from meta-analyses and extensive randomized controlled trial (RCT) studies [17,111]. The majority of the n-3LCPUFA supplementation trials were conducted during 16–20 weeks of gestation, well before the development of first-trimester pregnancy. Therefore, studies are required to evaluate the effects of n-3LCPUFA supplementation during pre-conception or before placentation prior to the 14th week of pregnancy to ensure optimal placentation and protect placental developmental-related abnormalities.

DHA incorporates a considerable amount of the fetal brain and retina during the third trimester of pregnancy [112,113], and this correlates with the development of normal eyesight and cognitive function [112]. DHA deficient condition can exist for an extended period after birth of very low birth weight (VLBW) babies since neonates miss the placenta's DHA supply window due to shortened gestation time. Various studies demonstrated the beneficial effects of maternal intakes of n-3 fats during pregnancy in optimum growth and development of the brain and retina [114]. Maternal supplementation of n-3 LCPUFAs during pregnancy increased gestational duration with a slight increase in birth weight [115,116]. The improvement in birth weight and prolong gestation was confirmed by meta-analyses of several n-3 LCPUFAs supplementation studies during pregnancy [115]. Despite these, the DHA supplementation had mixed impacts on gestation length in the general population of pregnant women [74,117]. Low consumption of marine fish is found a decisive risk factor for preterm delivery [118]. Daily intake of n-3PCPUFA significantly lowered the rate of recurrent preterm delivery as concluded from a large multicentric clinical trial involving subjects with a history of preterm delivery [115]. The effects of n-3LCPUFA supplementation in high-risk pregnancy showed an inconsistent outcome [117]. Regular intake of n-3PCPUFA or placebo failed to prevent the recurrence of high-risk IUGR or pregnancy-induced hypertension [119]. The power and sample size can determine the measurable differences in the n-3LCPUFA supplementation trial's birth outcome during pregnancy [17,113]. The improved birth weight due to n-3LCPUFA supplementation often arises due to the longer gestational length of these pregnancies [120]. The adequate intake of n-3 LCPUFA by the mother ensures increased availability of these fatty acids in the fetus [121]. Besides, post-natal supplementation of LCPUFAs also improved the subsequent development of vision and brain in VLBW infants [122].

2.3. Maternal Intakes of DHA and Its Impact on Fetal Brain Development

Despite studies supported the nutritional and metabolic requirement of n-3 PUFAs, but functional implications of the n-3 PUFA, like ALA, were not demonstrated. The first report on ALA deficiency was reported in 1982 when the parenteral formula containing a higher ALA (42.4% LA and 6.9% ALA) corrected the neurological disorders of a 6-years old girl [123]. Holman and coworkers then proposed that ALA was an essential fatty acid. The minimum requirement to prevent symptoms caused by ALA deficiency was in the range of 0.5–0.6% of total energy intake [123]. After consuming an ALA deficit diet, elderly patients developed dermatological disorders, particularly dermatitis, and flaky skin, together with deficient circulating EPA and DHA levels. The adverse skin symptoms were resolved, and plasma levels of EPA and DHA regained normal value once ALA was added to the formula [124]. Based on available evidence, a minimum intake of ALA and EPA plus DHA for an adult was fixed at 0.2–0.3 en% per day and 0.1–0.2 en% per day, respectively, indicating that in the absence of EPA and DHA, the endogenous biosynthesis of these fatty acids from ALA is significantly increased [125]. Considering data available to date and the substantial presence of DHA in the human nervous system, including in the brain and retina, it is agreed that humans can transform 1% of total ingested ALA into DHA [126]. The breast milk remained the only food supply to the neonate containing all the essential nutrients, including the necessary and obligatory n-3 and n-6 LCPUFAs, to ensure optimal brain development [127–129].

Colostrum levels of n-3 PUFAs are positively associated with infant mental development and [130]. In contrast, LA levels in colostrum are negatively associated with child cognitive scores at ages 2 and 3 years, independently of breastfeeding duration [131]. DHA's relative content is ranged from 0.1% to 1% of the total fatty acids in human breast milk. The DHA content is vastly varied among populations globally, depending on their eating habits of fish or seafood intake or food cycle with land animals that feed on fish-meal or fish oils [43,132]. Over time, the level of DHA in breast milk is significantly decreased in the Western population, primarily due to lower intake of DHA containing foods [133]. According to the Food and Agriculture Organization (FAO), minimum dietary intake of 200 mg/day is suggested for pregnant women [134]; however, there is no universal consensus on the recommendation. However, the compliance was much better when pregnant women were advised or received dietary counseling about DHA's physiological importance in pregnancy, as reflected in their increased consumption of n-3 LCPUFAs containing foods and/or supplements [135]. Dietary counseling about the benefits of marine fish or fish product consumption during pregnancy significantly boosted n-3 LCPUFAs [136]. FAO and many studies have established that benefit of consuming marine fish rich in n-3 LCPUFA quash the possible adverse effects of any heavy metal or other organic contaminants present [137]. Consumption of two portions of fish rich in DHA, such as salmon, tuna, anchovy, or mackerel, in a week may adequately comply with the minimum dietary requirements during pregnancy [138,139]. Low intake of marine foods that provide DHA during pregnancy is critical to obtain enough of these fatty acid levels, reflected in low DHA levels in umbilical blood [140]. Intake of *trans* fatty acid ingestion may also decrease the availability of n-3 LCPUFA to the mother and her child [141]. All these data suggest that the public health policies must be developed using DHA consumption for the population, especially pregnant and nursing women.

3. The DHA Uptake System in the Human Brain

Usually, non-esterified DHA is the major plasma pool for its supply to the brain [142,143]. DHA is also present as an esterified to lysophosphatidylcholine (LPC-DHA) pools in similar amounts. The brain maintains its fatty acid level via fatty acid uptake from circulation [144]. Astrocytes and endothelial cells of the blood-brain barrier (BBB) are significantly involved in the brain's uptake of plasma fatty acids. The brain's fatty acid uptake may occur via different mechanisms, such as passive diffusion and a saturable protein-mediated transport system.

In circulation, free fatty acids (FFAs) are mainly bound by albumin. At the inner surface of endothelial cell membranes, a small fraction of FFAs is delivered into the subcellular compartments for further metabolism. At the same time, most FFAs diffuse into the cytosol with or without the help of the battery of cell membrane- and cytoplasmic fatty acid-binding proteins. Figure 1 shows the putative fatty acid transport system of the brain. There are four classes of fatty acid transport proteins involved in transportation in the adult brain, including fatty acid translocase (FAT/CD36), plasma membrane-fatty acid-binding proteins (FABPpm), fatty acid transport proteins (FATPs), and cytoplasmic FABPs. Additionally, MFSD2a is newly identified as a DHA transporter in the brain. Even though there is a preferential uptake by the brain of esterified DHA (LPC-DHA), this is not the major pathway by which DHA is transported into the brain. Despite being an abundant fatty acid in brain phospholipids, DHA cannot be synthesized de novo in the brain and must be imported across the blood-brain barrier, but its transportation pathways are unknown yet. MFSD2a is the major DHA transporter, is expressed exclusively in the endothelium of the BBB of micro-vessels. The MFSD2a-deficient mice brain had significantly reduced DHA concentrations with neuronal cell loss in the hippocampus and cerebellum. These mice also exhibited had microcephaly with severe cognitive deficits and anxiety. MFSD2a transports LPC-DHA, but not unesterified DHA, in a sodium-dependent manner. Notably, MFSD2a transports plasma pool LPCs carrying long-chain fatty acids (C>14). Long-chain acyl-CoA synthetases (ACSLs) are also involved in brain DHA uptake [51,102]. Brain development requires an incredible increase in the de novo synthesis and accretion of DHA, mediated by several factors, including sterol regulatory element-binding protein SREBP. In normal physiology, the activity of MFSD2a is regulated by SREBP to maintain a balance between de novo lipogenesis and exogenous uptake of LPC-DHA [145].

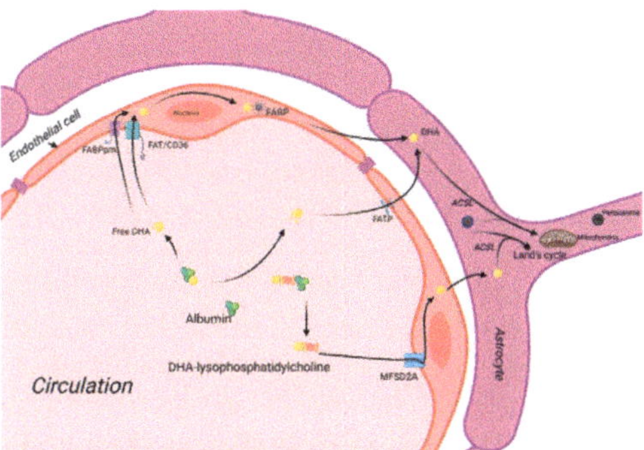

Figure 1. Docosahexaenoic acid (DHA) uptake system of the brain. DHA: Docosahexaenoic acid; MFSD2A: Major facilitator superfamily domain containing 2A; FABPpm: Plasma membrane fatty-acid binding proteins; FAT/CD36: Fatty acid translocase/cluster of differentiation 36; FABP: Fatty acid-binding protein; FATP: Fatty acid transporter protein; ACSL: long-chain acyl-CoA synthetase.

DHA is carried through the plasma by albumin and circulating lipoproteins. There are four classes of lipid transport proteins involved in DHA uptake of the brain that includes fatty acid translocase (FAT/CD36), plasma membrane fatty acid-transport proteins (FABPpm), and fatty acid transport proteins (FATPs) and cytoplasmic FABPs. MFSD2a, a specific protein, can transport plasma LPC-DHA, but not other DHA forms, across the blood-brain barrier to the neuron.

DHA Deficiency during Fetal Brain Development and Its Impact on Cognitive Functions

During pregnancy and lactation, maternal intake of DHA ensures adequate maternal reserve deposited during pregnancy to support six-month breast-fed post-natal life. Maternal adipose resources during pregnancy cover the increasing demand for DHA in the early post-natal stage. Although it is difficult to estimate the quantity of DHA require in the diet for optimum brain development, the study from Kuiper et al. [146] first estimated the absolute requirement of DHA, ARA, and LA at 25 (conceptual age), 35 (preterm), and 45 (term) weeks. Based on the mother's data fed on the western diet, the DHA accretion rate was found 42 mg/day in the last five weeks of pregnancy. The accretion of DHA was found double in the last five weeks of pregnancy compared with the first thirty-five weeks together. The DHA accretion rate largely determines the DHA deficiency during brain development in the brain during a term and preterm conditions. Thus, it is important to devise the DHA requirement for preterm babies based on the mother's data from a different ethnic and dietary background. The majority of the available data suggests DHA's optimal requirement during brain development in infants is fed on the Western diet. A scenario with high LA intake in the diet will further enhance the need for pre-formed DHA in infant formula. To fulfill enough maternal reserve of DHA, intake may be required well before conception.

In developing populations, where pregnant women's usual diet is low in fats in which n-6 PUFAs are predominantly present with little intake of ALA or DHA as n-3PUFAs, most of the women start pregnancy with inadequate or insufficient n-3 PUFA status in their reserve. Under such a scenario, where a maternal reserve of DHA is low, endogenous synthesis of DHA from precursors is inadequate, resulting in an insufficient supply of DHA to the fetus that may affect DHA accretion in the brain during the brain development phase of the neonate. DHA's placental deficiency can be correlated with the lower DHA accretion in the brain of the babies born preterm in the developing population. There is no clinical data available about DHA accretion in the brain and neonatal's cognitive performance from the mother fed on the n-3 deficient diet in such a population. Therefore, optimal maternal intake of n-3 LCPUFAs is essential to fulfilling DHA's neonatal requirement for the first six months. Maternal intake of DHA is correlated with the problem-solving skills of the children in some aspects. It was argued that inadequate intake of DHA could affect rapid accretion of DHA in the human brain, mainly when brain growth is maximal as with the third trimester or first six months of life. Adequate DHA during this period ensures the maturation of the prefrontal cortex's specific brain domain that may support the problem-solving skills.

The impact of DHA deficiencies on cognitive functions is extensively studied using animal models in vivo. DHA deficiency can influence DHA's endogenous synthesis in the brain and its impact on synaptic plasticity. A recent study with DHA deficient mice by silencing ELOVL2, an enzyme that is responsible for endogenous synthesis of DHA from its precursors, showed downregulation in the expression of neural plasticity factors (BDNF, Arc-1) with a concomitant increase in the pro-inflammatory markers (TNFα, IL-1β, inducible nitric oxide synthase iNOS) in the cerebral cortex of the DHA deficient mice [147]. The altered expression of these factors is also associated with the learning and memory function in the brain. The endogenous DHA deficient mice showed an alteration in microglial architecture and cytokine factors without involving astrocytes indicates that resident immune cells are affected in the brain by endogenous DHA deficiency. The replenishment with DHA restored the physiological expression of neuroinflammatory and neuroplasticity factors in the cerebral cortex. These data indicate that DHA plays a critical role in neuroimmune communication in brain function and synaptic plasticity.

The damaging effects of n-3 PUFA deficiency on brain lipid composition and memory performance were evidenced in lipopolysaccharides (LPS)-induced rat models, suggesting that in utero n-3 PUFAs deficiency could be a potential risk factor for neurodevelopmental disorders [148]. The n-3 PUFA deficiency in rats shows downregulated glutamate receptors and upregulated pro-inflammatory TNFα gene expression in the central nervous tissue independent of their effects on membrane composition [149]. Chronic deficiency of DHA over multiple generations during brain spurt affects the process of neurodevelopment by modulating the neuronal cell growth and differentiation, as well as neuronal signaling. The deficiency may cause a functional deficit in the offspring's learning and cognitive efficiency by reducing intellectual potential and enhancing the risks of neurological diseases in adult life [150]. The n-3 fatty acid deficiency disrupts the peripheral balance of pro-and anti-inflammatory states in the brain due to altered systemic ARA: DHA ratio [151]. The excess ARA generates high prostaglandin concentrations, leukotriene, and thromboxane that lead to a pro-inflammatory state in the brain that disrupts the balance of anti- (n-3) and pro-inflammatory (n-6) eicosanoids due to alteration in GPR receptors' signaling [152]. In the animal model, DHA's maternal deficiency revealed reduced telencephalon structure in the hippocampus region [153]. Such a deficiency state affects region-specific brain development areas where the cerebral frontal cortex region is affected mostly, leading to hyper motor activity, reduced learning ability, and altered monoamine transmission [154]. The maternal DHA deficiency affects the offspring's brain development in a gender-specific manner due to differential efficiency of endogenous DHA converting enzyme in males and females.

As the endogenous conversion of DHA from its precursor is more efficient in female newborns due to estrogen presence, the infant male is more susceptible to the risk of brain disorders, such as ADHD, Autism, etc., in their later life [155]. The maternal DHA deficiency state profoundly affects behavioral change in feed intake, anxiety, and stress response in the offspring [156]. The n-3 deficiency state triggers sucrose-motivated food intake preference due to alteration in the brain-rewarding pathway that may prompt children to consume calorie-dense foods [157]. Chronic deficiency of n-3 fatty acids for multiple generations induces anxiety-related stress behavior in the offspring due to altered expression of neuropeptide Y-1 receptor and glucocorticoid receptor in the pre-frontal and hippocampus of the rat brain [150,158]. Data from several in vivo studies suggest that DHA promotes neurogenesis by improving the membrane fluidity in the structural domain of the hippocampus, prefrontal cortex, and hypothalamus region to stabilize the neurodevelopment circuitry network required for learning and memory recognition processes [159]. N-3 fatty acid deficiency during utero development and in the postnatal state negatively impacts cognitive abilities [74,160]. DHA's dietary deficiency increases the risk for neurocognitive disorders, whereas a diet enriched with DHA increases learning and memory and protects against cognitive decline during aging. However, whether increased intake of DHA can prevent the risk of brain disorders requires further investigation.

4. Maternal DHA Supplementation and Brain Development

During pregnancy, DHA's importance for fetal brain development has been shown in a large observational study (N = 11,875). The study found that children born to mothers with a higher intake of seafood during pregnancy improved fine motor skills, more significant pro-social behavior, higher verbal intelligence, and higher social development scores at eight years of age [66]. A recent longitudinal cohort concludes that higher DHA status during pregnancy and lactation is associated with an infant's problem-solving skills at 12 months [161]. But Crozier did not find any relationship between DHA level during pregnancy and cognitive performance of 4- or 6-year-old children [161].

A randomized controlled trial (RCT) found that taking 200 mg DHA orally daily for 4 months after delivery caused children's higher cognitive abilities at 5 years of age [162]. Ogundipe et al. showed that 300 mg/day DHA supplementation on the last trimester of pregnancy correlates positively with an infant's brain volumes (on MRI scan) [163]. DHA (600 mg/day) supplementation from 14.5 weeks of pregnancy until the delivery on KUDOS study found improved visual attention in infancy but no consistent long-term benefit in childhood [164]. DHA (120 mg/day) supplementation, along with EPA (180 mg/day), from the 20th week of pregnancy until the 30th day of the postpartum period among 18–35 years older women in Iran found primary neurodevelopment improvement among 4 to 6 months old children [165]. However, another RCT from Australia did not find any effect among the toddlers at 18 months of age after an 800 mg DHA intake in 2399 pregnant women during <21st weeks of gestation to delivery [17]. Maternal supplementation of DHA (400 mg/day) during pregnancy positively reflected the child's (5–6 years) performance on language skills and short-term memory [166]. The positive effect of maternal DHA intake during gestation was observed in 18-month-old infants but not when they were 5 years old. DHA's long-term effects may be too small to detect, or it is possible that the DHA intake was not good enough during gestation to have lasting effects.

A considerable DHA accretion occurs during the brain growth spurt beginning in the third trimester. However, there is no adequate information on whether DHA deficiency during early gestation can impact fetal brain development in the third trimester. Most DHA intervention studies on neurodevelopment were carried out from the second trimester of pregnancy. Therefore, the impacts of DHA intervention at 16 weeks of gestation may be too late as DHA is required during the first trimester for early placental growth and development, the critical step for future placenta's adequate roles for the maternal supply of DHA for the fetal brain development. The inverse relationship of child Beery scores with maternal erythrocytes 22:4n-6 and 22: 5n-6 suggests that visual-motor integration development is compromised at low prenatal DHA. This is consistent with the time course of brain maturation; maturation occurs in the visual cortex before the prefrontal cortex.

Consuming DHA-rich eggs (135 mg DHA/egg) during pregnancy showed higher erythrocyte and umbilical cord DHA levels as compared with those pregnant women who consumed non-enriched-eggs (18 mg DHA/egg) [167]. Babies born to mothers who consumed DHA-enriched cereal-bars (300 mg DHA/bar) had increased visual acuity until four months of postpartum [168], and a better capacity to resolve problems with an improved organization of their dream [169]. Despite the beneficial effects of fish oil n-3 fatty acids in fetal development, recommendations to increase fish consumption for pregnant women are often met by the fear of heavy metal contaminations in these foods. However, many health professionals recommend avoiding or reducing fish consumption, especially for pregnant women. Despite some organoleptic problems, because the reconstituted product developed some uncomfortable smell that is not accepted for some mothers, the formula has proved to increase the DHA content of breast milk [170]. In Chile, a newly developed inexpensive formula overcame this organoleptic limitation and increased DHA availability for pregnant and nursing population [171].

Available data based on randomized controlled trials suggest beneficial effects of maternal supplementation of DHA on neonatal growth and cognitive development (Table 1). Several RCTs found that DHA supplementation on term and pre-term infants found a significant outcome on visual development during infancy [172]. Table 2 presents RCTs from the postnatal supplementation of DHA on the neonatal visual, verbal, and cognitive development. A meta-analysis of RCTs on routinely supplemented infant formula milk with DHA has found no beneficial role in neurodevelopmental outcomes [173]. A large trial (DHANI) was carried out in India, where prenatal and 6 months of post-partum 400 mg/day maternal DHA supplementation effects measure to evaluate the neurodevelopment outcome [174,175]. The study reported that the mean development quotient (DQ) scores in the DHA and placebo groups were not statistically significant after 12 months of mothers' supplementation through pregnancy and lactation with 400 mg/d DHA [175].

Table 1. Supplementation of DHA on the neonatal growth and cognitive development: a consolidated Randomized Controlled Trial (RCT).

Subject, Sample Size, Location	Dosages, Duration	Primary Outcome	References
Pregnant women, $n = 350$, USA	DHA 600 mg per day, <20 wk to delivery	Gestational duration ↑ Birth size ↑	Carlson et al., 2013 [45]
Pregnant women, $n = 315$, Germany, and others	DHA 500 mg and EPA 150 mg per day, <20 wk to delivery	Visual coordination 2.5 yr. children ↑ Cognitive development 5.5 yr children↑	Dunstan et al., 2008 Escolano et al., 2011 [52]
Pregnant women, $n = 300$, UK	DHA 300 mg, EPA 42 mg, ARA 8.4 mg per day for 12 wks from the third trimester	MRI of infant ($n = 86$) at birth show a correlation with DHA and brain volume ↑	Ogundipe et al., 2018 [163]
Pregnant women, $n = 271$, Canada	DHA 400 mg per day, 16 wk to delivery	Maternal DHA correlates with language and short-term memory development of 5.79 yr children ↑	Mulder et al., 2018 [166]
Pregnant women, $n = 1094$, Mexico	DHA 400 mg per day, 18–22 wk to delivery	Birth size and head circumference at birth ↑ The attention of 5 yr pre-school children↑	Ramkrishnan et al., 2010 [113]
Pregnant women, $n = 2399$, Australia	DHA 800 mg per day, <21 wk to delivery	No effects on cognitive and language development in 1.2 yr infant	Makrides et al., 2010 [17]
Pregnant women, $n = 301$, USA	DHA 600 mg per day, 14.5 wk to delivery	Cognitive behavior 10 mo to 6 yr; Visual attention ↑ No long-term beneficial effects	Colombo et al., 2019 [164]
Pregnant women, $n = 143$, Norway	DHA 1183 mg and EPA 803 mg per 10 mL per day, 18 wk to post-delivery 3 mo	Mental processing score at 4 and 7 yr age ↑ No effects on BMI at 7 yr age	Helland et al., 2003 [65]
Pregnant women, $n = 150$, Iran	DHA 120 mg and EPA 180 mg per day, 20 wk to post-delivery 1 mo.	Primary neurodevelopment outcome of 4–6 mo Infant ↑	Ostadrahim et al., 2018 [165]
Pregnant women, $n = 98$, Australia	DHA 2200 mg and EPA 1100 mg per day, 20 wk to delivery	Visual and coordination 2.5 yr children ↑	Dunstan et al., 2008 [52]
Pregnant women, $n = 30$, USA	DHA 214 mg as a functional food, 24 wk to delivery	Visual acuity 4 mo infant ↑	Judge et al., 2007 [168]

Abbreviations: DHA: Docosahexaenoic acid; ARA: Arachidonic acid; EPA: Eicosapentaenoic acid; Mo: Month; Yr: Year; Wk: Week; MRI: Magnetic resonance imaging. ↑ denotes "increased".

Table 2. Supplementation of DHA on the neonatal visual, verbal, and cognitive development: a collection of Randomized Controlled Trials (RCTs).

Subject, Sample Size, Location	Dosages, Duration	Measured Outcome	References
Term formula-fed infant, $n = 343$, USA	DHA (0.32%–0.96%), ARA (0.64%) from 1–9 day to 1 yr	DHA (0.32%) group visual acuity ↑	Birch et al., 2010 [73]
Term infant, $n = 420$, Australia	DHA 250 mg and EPA 60 mg per day, birth to 6 mo	Accretion of DHA ↑ Early development of language and communication skills ↑	Meldrum et al., 2012 [76]
Infant BW < 1.5 kg, $n = 141$, Norway	DHA 32 mg, ARA 31 mg per 100 mL human milk per day, 1 to 9 wk after birth	Memory recognition and problem-solving skills of 6 mo infant ↑	Henriksen et al., 2008 [74]
Pre-term infant, $n = 107$, UK	DHA (0.5%) in the supplemented formula from birth to 9 mo	Verbal and intellectual coefficient of 9 yr girl ↑	Isaacs et al., 2011 [77]
Pre-school healthy children, $n = 175$, USA	DHA 400 mg per day for 4 mo of 4-yr-old children	Higher blood DHA correlates comprehension, punctuation, and vocabulary abilities ↑	Ryan et al., 2008 [85]
Indigenous school children three to 13 yr, $n = 409$, Australia	DHA 750 mg, GLA 60 mg per day for 20 school week	Scholar performance in 7–12 yr children ↑	Parletta et al., 2013 [86]
Term infant, $n = 227$, USA	Algal DHA 200 mg per day for 4 mo after delivery	Cognitive abilities in 5 yr children ↑	Jensen et al., 2010 [162]
Pre-term infant, $n = 361$, USA	Algal DHA oil (17 mg/100 kcal), fungal ARA oil (34 mg/100 mL) from birth to 4 mo	Growth and development of pre-term infant till 118 wk ↑	Clandinin et al., 2005 [176]
Pre-term 23–24 wk infant, $n = 90$, USA	Algal DHA 50 mg per day from 1 to 6–7 wk at discharge	DHA levels of pre-term infant comparable to term placebo ↑	Baack et al, 2016 [177]
ADHD 7–12 yr children, $n = 90$, Australia	DHA 1032 mg or 108 mg, EPA 264 mg or 1109 mg for 4 mo	Reading and spelling correlated DHA levels in the blood ↑	Milte et al, 2012 [87]
Larger pre-term 30–37 wk, BW > 2 kg, $n = 27$, Taiwan	DHA (0.05%), ARA (0.1%) infant formula for 6 mo after birth	Mental development index at 6–12 mo ↑	Fang et al, 2005 [75]
Breastfed 6 mo infant, $n = 25$–26, USA	DHA 115 mg/100 g baby food from 6 mo to 12 mo of the breastfed infant	Maturation of retina and visual cortex at 12 mo ↑	Hoffman et al., 204 [69]

↑ denotes "increased".

Figure 2 shows the accretion of DHA at different development stages that affect fetoplacental and fetal brain development. DHA accumulates substantially in the retina and cerebral cortex during the last trimester and the second year of life. Intervention studies have shown that improving maternal DHA nutrition reduces the risk of visual and neural development in infants and children. Several pieces of evidence supports the notion that maternal transfer of DHA to the infant before and after birth, with short and long-term, modulates neural functions. However, genetic variation responsible for endogenous conversion of DHA by fatty acid desaturases also influences essential fatty acid metabolism and may affect individuals' optimal DHA requirements. Consideration of adequate DHA intake includes brain development, a balanced intake of n-3, and n-6 PUFAs in gestation, and lactation, and optimal fatty acid nutrition during pregnancy is required for infant neurodevelopment. Premature infants have a deficit in DHA shortly after delivery for several reasons, including missing the third trimester DHA accretion. More studies are required to assess DHA's optimal dosage, delivery method, and duration of supplementation to evaluate DHA intake in premature infants.

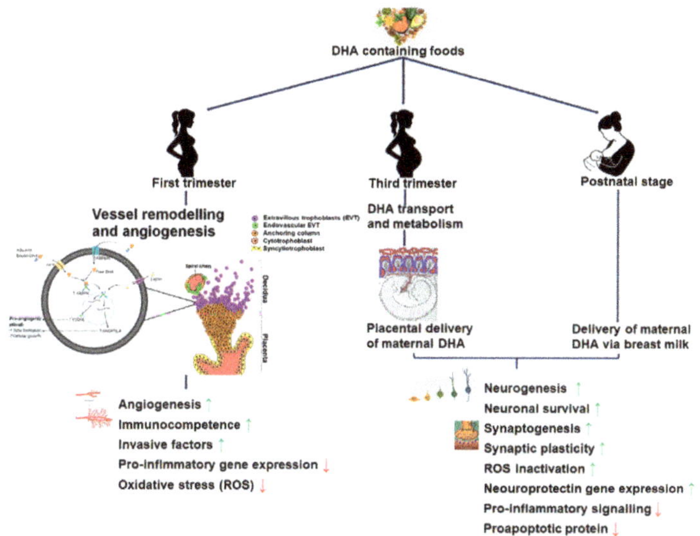

Figure 2. Maternal source and delivery of DHA at different stages of development affect fetoplacental and fetal brain development.

DHA plays different roles in the gestation specific development of the feto-placental unit. During the first trimester, DHA possibly involves in decidual remodeling to establish early placentation by promoting the expression and secretion of angiogenic growth factors, such as VEGFA, FABP4, ANGPTL4, leptin, etc. DHA and its metabolites improve maternal-fetal immunocompetence by maintaining the oxidative stress, production of reactive oxygen species (ROS), and pro-anti-inflammatory balance maternal-fetal interface. DHA is delivered to the fetal brain during the third trimester via the placenta and subsequently via breast milk. In addition to DHA's presence in the neonatal brain matter's structural skeleton, mounting evidence suggests DHA helps several brain development processes, including neurogenesis, synaptogenesis, brain plasticity, inflammatory signaling, neuroprotection, etc.

5. Conclusions

The relationships between DHA on the brain development and function have been extensively studied. Under its ability to control membrane fluidity, the DHA also modulates neuronal density, neurotransmitter concentration, and synaptic activity by regulating the brain's neuroinflammatory state. Increasing evidence suggests that DHA, the foremost important n-3 long-chain fatty acid in the brain, has neurotrophic and neuroprotective properties. Dietary n-3 PUFA deficiency during early development decreased the brain n-3 PUFA levels. Together, this body of evidence supports the proposition that DHA deficiency in utero increases the vulnerability of brain development. Most evidence indicates that the DHA accumulation is mainly influenced by dietary intake, specifically of preformed DHA. Insufficient intake of n-3 PUFA may lead to DHA deficiency states that could affect the offspring's metabolic phenotypes by altering placental structure and function, fetal adiposity, body fat distribution, energy utilization, musculoskeletal growth, signaling between brain-adipose, epigenome stability, and inflammation [10].

Author Contributions: Conceptualization, A.K.D.; Methodology, A.K.D. and S.B.; Investigation, A.K.D. and S.B. Writing—Original Draft Preparation A.K.D.; Writing—Review & Editing, A.K.D., S.B., and R.M.; Visualization, R.M., S.B., and A.K.D.; Supervision, A.K.D. All authors have read and agreed to the published version of the manuscript.

Funding: This research received no external funding.

Acknowledgments: This work is supported in part by the grants from Throne Holst Foundation, and the Faculty of Medicine, University Oslo, Norway.

Conflicts of Interest: Authors express no conflict of interest.

References

1. Salem, N., Jr.; Litman, B.; Kim, H.Y.; Gawrisch, K. Mechanisms of action of docosahexaenoic acid in the nervous system. *Lipids* **2001**, *36*, 945–959. [CrossRef] [PubMed]
2. Bourre, J.M. Dietary omega-3 fatty acids for women. *Biomed. Pharm.* **2007**, *61*, 105–112. [CrossRef] [PubMed]
3. Bradbury, J. Docosahexaenoic acid (DHA): An ancient nutrient for the modern human brain. *Nutrients* **2011**, *3*, 529–554. [CrossRef]
4. Horrocks, L.A.; Yeo, Y.K. Health benefits of docosahexaenoic acid (DHA). *Pharmacol. Res.* **1999**, *40*, 211–225. [CrossRef]
5. Quinn, J.F.; Raman, R.; Thomas, R.G.; Yurko-Mauro, K.; Nelson, E.B.; Van Dyck, C.; Galvin, J.E.; Emond, J.; Jack, C.R., Jr.; Weiner, M.; et al. Docosahexaenoic acid supplementation and cognitive decline in Alzheimer disease: A randomized trial. *J. Am. Med. Assoc.* **2010**, *304*, 1903–1911. [CrossRef]
6. de Urquiza, A.M.; Liu, S.; Sjoberg, M.; Zetterstrom, R.H.; Griffiths, W.; Sjovall, J.; Perlmann, T. Docosahexaenoic acid, a ligand for the retinoid X receptor in mouse brain. *Science* **2000**, *290*, 2140–2144. [CrossRef]
7. Ji, R.R.; Xu, Z.Z.; Strichartz, G.; Serhan, C.N. Emerging roles of resolvins in the resolution of inflammation and pain. *Trends Neurosci.* **2011**, *34*, 599–609. [CrossRef]
8. Matesanz, N.; Park, G.; McAllister, H.; Leahey, W.; Devine, A.; McVeigh, G.E.; Gardiner, T.A.; McDonald, D.M. Docosahexaenoic acid improves the nitroso-redox balance and reduces VEGF-mediated angiogenic signaling in microvascular endothelial cells. *Invest. Ophthalmol. Vis. Sci.* **2010**, *51*, 6815–6825. [CrossRef] [PubMed]
9. Kabaran, S.; Besler, H.T. Do fatty acids affect fetal programming? *J. Health Popul. Nutr.* **2015**, *33*, 14. [CrossRef] [PubMed]
10. Basak, S.; Vilasagaram, S.; Duttaroy, A.K. Maternal dietary deficiency of n-3 fatty acids affects metabolic and epigenetic phenotypes of the developing fetus. *Prostaglandins Leukot. Essent. Fatty Acids* **2020**, *158*, 102109. [CrossRef]
11. Lacombe, R.J.S.; Chouinard-Watkins, R.; Bazinet, R.P. Brain docosahexaenoic acid uptake and metabolism. *Mol. Asp. Med.* **2018**, *64*, 109–134. [CrossRef] [PubMed]
12. Brenna, J.T.; Carlson, S.E. Docosahexaenoic acid and human brain development: Evidence that a dietary supply is needed for optimal development. *J. Hum. Evol.* **2014**, *77*, 99–106. [CrossRef] [PubMed]
13. Niemoller, T.D.; Bazan, N.G. Docosahexaenoic acid neurolipidomics. *Prostaglandins Other Lipid Mediat.* **2010**, *91*, 85–89. [CrossRef] [PubMed]
14. Salem, N., Jr.; Eggersdorfer, M. Is the world supply of omega-3 fatty acids adequate for optimal human nutrition? *Curr. Opin. Clin. Nutr. Metab. Care* **2015**, *18*, 147–154. [CrossRef]
15. Buckley, C.D.; Gilroy, D.W.; Serhan, C.N. Proresolving lipid mediators and mechanisms in the resolution of acute inflammation. *Immunity* **2014**, *40*, 315–327. [CrossRef]
16. Serhan, C.N.; Dalli, J.; Colas, R.A.; Winkler, J.W.; Chiang, N. Protectins and maresins: New pro-resolving families of mediators in acute inflammation and resolution bioactive metabolome. *Biochim. Biophys. Acta* **2015**, *1851*, 397–413. [CrossRef]
17. Makrides, M.; Gibson, R.A.; McPhee, A.J.; Yelland, L.; Quinlivan, J.; Ryan, P.; Team, D.O.I. Effect of DHA supplementation during pregnancy on maternal depression and neurodevelopment of young children: A randomized controlled trial. *JAMA* **2010**, *304*, 1675–1683. [CrossRef]

18. Funk, C.D. Prostaglandins and leukotrienes: Advances in eicosanoid biology. *Science* **2001**, *294*, 1871–1875. [CrossRef]

19. Latham, C.F.; Osborne, S.L.; Cryle, M.J.; Meunier, F.A. Arachidonic acid potentiates exocytosis and allows neuronal SNARE complex to interact with Munc18a. *J. Neurochem.* **2007**, *100*, 1543–1554. [CrossRef]

20. Wu, A.; Ying, Z.; Gomez-Pinilla, F. The salutary effects of DHA dietary supplementation on cognition, neuroplasticity, and membrane homeostasis after brain trauma. *J. Neurotrauma* **2011**, *28*, 2113–2122. [CrossRef]

21. Belayev, L.; Khoutorova, L.; Atkins, K.D.; Eady, T.N.; Hong, S.; Lu, Y.; Obenaus, A.; Bazan, N.G. Docosahexaenoic Acid therapy of experimental ischemic stroke. *Transl. Stroke Res.* **2011**, *2*, 33–41. [CrossRef] [PubMed]

22. Arita, M.; Bianchini, F.; Aliberti, J.; Sher, A.; Chiang, N.; Hong, S.; Yang, R.; Petasis, N.A.; Serhan, C.N. Stereochemical assignment, antiinflammatory properties, and receptor for the omega-3 lipid mediator resolvin E1. *J. Exp. Med.* **2005**, *201*, 713–722. [CrossRef] [PubMed]

23. Marcheselli, V.L.; Hong, S.; Lukiw, W.J.; Tian, X.H.; Gronert, K.; Musto, A.; Hardy, M.; Gimenez, J.M.; Chiang, N.; Serhan, C.N.; et al. Novel docosanoids inhibit brain ischemia-reperfusion-mediated leukocyte infiltration and pro-inflammatory gene expression. *J. Biol. Chem.* **2003**, *278*, 43807–43817. [CrossRef]

24. Gronert, K.; Maheshwari, N.; Khan, N.; Hassan, I.R.; Dunn, M.; Laniado Schwartzman, M. A role for the mouse 12/15-lipoxygenase pathway in promoting epithelial wound healing and host defense. *J. Biol. Chem.* **2005**, *280*, 15267–15278. [CrossRef]

25. Serhan, C.N.; Gotlinger, K.; Hong, S.; Lu, Y.; Siegelman, J.; Baer, T.; Yang, R.; Colgan, S.P.; Petasis, N.A. Anti-inflammatory actions of neuroprotectin D1/protectin D1 and its natural stereoisomers: Assignments of dihydroxy-containing docosatrienes. *J. Immunol.* **2006**, *176*, 1848–1859. [CrossRef]

26. Mukherjee, P.K.; Marcheselli, V.L.; Serhan, C.N.; Bazan, N.G. Neuroprotectin D1: A docosahexaenoic acid-derived docosatriene protects human retinal pigment epithelial cells from oxidative stress. *Proc. Natl. Acad. Sci. USA* **2004**, *101*, 8491–8496. [CrossRef]

27. Chen, P.; Vericel, E.; Lagarde, M.; Guichardant, M. Poxytrins, a class of oxygenated products from polyunsaturated fatty acids, potently inhibit blood platelet aggregation. *FASEB J. Off. Publ. Fed. Am. Soc. Exp. Biol.* **2011**, *25*, 382–388. [CrossRef]

28. Liu, M.; Boussetta, T.; Makni-Maalej, K.; Fay, M.; Driss, F.; El-Benna, J.; Lagarde, M.; Guichardant, M. Protectin DX, a double lipoxygenase product of DHA, inhibits both ROS production in human neutrophils and cyclooxygenase activities. *Lipids* **2014**, *49*, 49–57. [CrossRef]

29. White, P.J.; St-Pierre, P.; Charbonneau, A.; Mitchell, P.L.; St-Amand, E.; Marcotte, B.; Marette, A. Protectin DX alleviates insulin resistance by activating a myokine-liver glucoregulatory axis. *Nat. Med.* **2014**, *20*, 664–669. [CrossRef]

30. Stein, K.; Stoffels, M.; Lysson, M.; Schneiker, B.; Dewald, O.; Kronke, G.; Kalff, J.C.; Wehner, S. A role for 12/15-lipoxygenase-derived proresolving mediators in postoperative ileus: Protectin DX-regulated neutrophil extravasation. *J. Leukoc. Biol.* **2015**, *99*, 231–239. [CrossRef]

31. Serhan, C.N.; Yang, R.; Martinod, K.; Kasuga, K.; Pillai, P.S.; Porter, T.F.; Oh, S.F.; Spite, M. Maresins: Novel macrophage mediators with potent antiinflammatory and proresolving actions. *J. Exp. Med.* **2009**, *206*, 15–23. [CrossRef] [PubMed]

32. Abdulnour, R.E.; Dalli, J.; Colby, J.K.; Krishnamoorthy, N.; Timmons, J.Y.; Tan, S.H.; Colas, R.A.; Petasis, N.A.; Serhan, C.N.; Levy, B.D. Maresin 1 biosynthesis during platelet-neutrophil interactions is organ-protective. *Proc. Natl. Acad. Sci. USA* **2014**, *111*, 16526–16531. [CrossRef] [PubMed]

33. Nordgren, T.M.; Heires, A.J.; Wyatt, T.A.; Poole, J.A.; LeVan, T.D.; Cerutis, D.R.; Romberger, D.J. Maresin-1 reduces the pro-inflammatory response of bronchial epithelial cells to organic dust. *Respir. Res.* **2013**, *14*, 51. [CrossRef] [PubMed]

34. Poorani, R.; Bhatt, A.N.; Dwarakanath, B.S.; Das, U.N. COX-2, aspirin and metabolism of arachidonic, eicosapentaenoic and docosahexaenoic acids and their physiological and clinical significance. *Eur. J. Pharmacol.* **2015**, *785*, 116–132. [CrossRef] [PubMed]

35. Gurzell, E.A.; Teague, H.; Harris, M.; Clinthorne, J.; Shaikh, S.R.; Fenton, J.I. DHA-enriched fish oil targets B cell lipid microdomains and enhances ex vivo and in vivo B cell function. *J. Leukoc. Biol.* **2013**, *93*, 463–470. [CrossRef] [PubMed]

36. Anderson, G.J.; Connor, W.E.; Corliss, J.D. Docosahexaenoic acid is the preferred dietary n-3 fatty acid for the development of the brain and retina. *Pediatric Res.* **1990**, *27*, 89–97. [CrossRef]

37. O'Brien, J.S.; Sampson, E.L. Fatty acid and fatty aldehyde composition of the major brain lipids in normal human gray matter, white matter, and myelin. *J. Lipid Res.* **1965**, *6*, 545–551.
38. Brenna, J.T.; Diau, G.Y. The influence of dietary docosahexaenoic acid and arachidonic acid on central nervous system polyunsaturated fatty acid composition. *Prostaglandins Leukot. Essent. Fat. Acids* **2007**, *77*, 247–250. [CrossRef]
39. Innis, S.M. Chapter 10 Essential fatty acid metabolism during early development. *Biol. Grow. Anim.* **2005**, *3*, 235–274. [CrossRef]
40. Campoy, C.; Escolano-Margarit, M.V.; Anjos, T.; Szajewska, H.; Uauy, R. Omega 3 fatty acids on child growth, visual acuity and neurodevelopment. *Br. J. Nutr.* **2012**, *107*, S85–S106. [CrossRef]
41. Montgomery, C.; Speake, B.K.; Cameron, A.; Sattar, N.; Weaver, L.T. Maternal docosahexaenoic acid supplementation and fetal accretion. *Br. J. Nutr.* **2003**, *90*, 135–145. [CrossRef] [PubMed]
42. McCann, J.C.; Ames, B.N. Is docosahexaenoic acid, an n-3 long-chain polyunsaturated fatty acid, required for development of normal brain function? An overview of evidence from cognitive and behavioral tests in humans and animals. *Am. J. Clin. Nutr.* **2005**, *82*, 281–295. [CrossRef] [PubMed]
43. Innis, S.M. Perinatal biochemistry and physiology of long-chain polyunsaturated fatty acids. *J. Pediatrics* **2003**, *143*, S1–S8. [CrossRef]
44. Helland, I.B.; Saugstad, O.D.; Smith, L.; Saarem, K.; Solvoll, K.; Ganes, T.; Drevon, C.A. Similar effects on infants of n-3 and n-6 fatty acids supplementation to pregnant and lactating women. *Pediatrics* **2001**, *108*, E82. [CrossRef]
45. Carlson, S.E.; Colombo, J.; Gajewski, B.J.; Gustafson, K.M.; Mundy, D.; Yeast, J.; Georgieff, M.K.; Markley, L.A.; Kerling, E.H.; Shaddy, D.J. DHA supplementation and pregnancy outcomes. *Am. J. Clin. Nutr.* **2013**, *97*, 808–815. [CrossRef] [PubMed]
46. Decsi, T.; Campoy, C.; Koletzko, B. Effect of N-3 polyunsaturated fatty acid supplementation in pregnancy: The Nuheal trial. *Adv. Exp. Med. Biol.* **2005**, *569*, 109–113.
47. Imhoff-Kunsch, B.; Stein, A.D.; Villalpando, S.; Martorell, R.; Ramakrishnan, U. Docosahexaenoic acid supplementation from mid-pregnancy to parturition influenced breast milk fatty acid concentrations at 1 month postpartum in Mexican women. *J. Nutr.* **2011**, *141*, 321–326. [CrossRef]
48. Helland, I.B.; Saugstad, O.D.; Saarem, K.; Van Houwelingen, A.C.; Nylander, G.; Drevon, C.A. Supplementation of n-3 fatty acids during pregnancy and lactation reduces maternal plasma lipid levels and provides DHA to the infants. *J. Matern. Fetal Neonatal Med.* **2006**, *19*, 397–406. [CrossRef]
49. Bergmann, R.L.; Haschke-Becher, E.; Klassen-Wigger, P.; Bergmann, K.E.; Richter, R.; Dudenhausen, J.W.; Grathwohl, D.; Haschke, F. Supplementation with 200 mg/day docosahexaenoic acid from mid-pregnancy through lactation improves the docosahexaenoic acid status of mothers with a habitually low fish intake and of their infants. *Ann. Nutr. Metab.* **2008**, *52*, 157–166. [CrossRef]
50. Larque, E.; Krauss-Etschmann, S.; Campoy, C.; Hartl, D.; Linde, J.; Klingler, M.; Demmelmair, H.; Cano, A.; Gil, A.; Bondy, B.; et al. Docosahexaenoic acid supply in pregnancy affects placental expression of fatty acid transport proteins. *Am. J. Clin. Nutr.* **2006**, *84*, 853–861. [CrossRef]
51. Duttaroy, A.K. Transport of fatty acids across the human placenta: A review. *Prog. Lipid Res.* **2009**, *48*, 52–61. [CrossRef] [PubMed]
52. Dunstan, J.A.; Simmer, K.; Dixon, G.; Prescott, S.L. Cognitive assessment of children at age 2(1/2) years after maternal fish oil supplementation in pregnancy: A randomised controlled trial. *Arch. Dis. Child. Fetal Neonatal Ed.* **2008**, *93*, F45–F50. [CrossRef] [PubMed]
53. Escolano-Margarit, M.V.; Ramos, R.; Beyer, J.; Csabi, G.; Parrilla-Roure, M.; Cruz, F.; Perez-Garcia, M.; Hadders-Algra, M.; Gil, A.; Decsi, T.; et al. Prenatal DHA status and neurological outcome in children at age 5.5 years are positively associated. *J. Nutr.* **2011**, *141*, 1216–1223. [CrossRef] [PubMed]
54. Campoy, C.; Escolano-Margarit, M.V.; Ramos, R.; Parrilla-Roure, M.; Csabi, G.; Beyer, J.; Ramirez-Tortosa, M.C.; Molloy, A.M.; Decsi, T.; Koletzko, B.V. Effects of prenatal fish-oil and 5-methyltetrahydrofolate supplementation on cognitive development of children at 6.5 y of age. *Am. J. Clin. Nutr.* **2011**, *94*, 1880S–1888S. [CrossRef]
55. Sanders, T.A.; Naismith, D.J. A comparison of the influence of breast-feeding and bottle-feeding on the fatty acid composition of the erythrocytes. *Br. J. Nutr.* **1979**, *41*, 619–623. [CrossRef]
56. Clandinin, M.T.; Chappell, J.E.; Leong, S.; Heim, T.; Swyer, P.R.; Chance, G.W. Extrauterine fatty acid accretion in infant brain: Implications for fatty acid requirements. *Early Hum. Dev.* **1980**, *4*, 131–138. [CrossRef]

57. Makrides, M.; Neumann, M.A.; Gibson, R.A. Perinatal characteristics may influence the outcome of visual acuity. *Lipids* **2001**, *36*, 897–900. [CrossRef]

58. Innis, S.M.; Auestad, N.; Siegman, J.S. Blood lipid docosahexaenoic and arachidonic acid in term gestation infants fed formulas with high docosahexaenoic acid, low eicosapentaenoic acid fish oil. *Lipids* **1996**, *31*, 617–625. [CrossRef]

59. Sanders, T.A.; Ellis, F.R.; Dickerson, J.W. Studies of vegans: The fatty acid composition of plasma choline phosphoglycerides, erythrocytes, adipose tissue, and breast milk, and some indicators of susceptibility to ischemic heart disease in vegans and omnivore controls. *Am. J. Clin. Nutr.* **1978**, *31*, 805–813. [CrossRef]

60. Innis, S.M.; Gilley, J.; Werker, J. Are human milk long-chain polyunsaturated fatty acids related to visual and neural development in breast-fed term infants? *J. Pediatrics* **2001**, *139*, 532–538. [CrossRef]

61. Lauritzen, L.; Brambilla, P.; Mazzocchi, A.; Harslof, L.B.; Ciappolino, V.; Agostoni, C. DHA Effects in Brain Development and Function. *Nutrients* **2016**, *8*, 6. [CrossRef] [PubMed]

62. Colombo, J.; Kannass, K.N.; Shaddy, D.J.; Kundurthi, S.; Maikranz, J.M.; Anderson, C.J.; Blaga, O.M.; Carlson, S.E. Maternal DHA and the development of attention in infancy and toddlerhood. *Child. Dev.* **2004**, *75*, 1254–1267. [CrossRef] [PubMed]

63. Malcolm, C.A.; McCulloch, D.L.; Montgomery, C.; Shepherd, A.; Weaver, L.T. Maternal docosahexaenoic acid supplementation during pregnancy and visual evoked potential development in term infants: A double blind, prospective, randomised trial. *Arch. Dis. Child. Fetal Neonatal Ed.* **2003**, *88*, F383–F390. [CrossRef] [PubMed]

64. Lauritzen, L.; Jorgensen, M.H.; Mikkelsen, T.B.; Skovgaard l, M.; Straarup, E.M.; Olsen, S.F.; Hoy, C.E.; Michaelsen, K.F. Maternal fish oil supplementation in lactation: Effect on visual acuity and n-3 fatty acid content of infant erythrocytes. *Lipids* **2004**, *39*, 195–206. [CrossRef] [PubMed]

65. Helland, I.B.; Smith, L.; Saarem, K.; Saugstad, O.D.; Drevon, C.A. Maternal supplementation with very-long-chain n-3 fatty acids during pregnancy and lactation augments children's IQ at 4 years of age. *Pediatrics* **2003**, *111*, e39–e44. [CrossRef] [PubMed]

66. Hibbeln, J.R.; Davis, J.M.; Steer, C.; Emmett, P.; Rogers, I.; Williams, C.; Golding, J. Maternal seafood consumption in pregnancy and neurodevelopmental outcomes in childhood (ALSPAC study): An observational cohort study. *Lancet* **2007**, *369*, 578–585. [CrossRef]

67. Oken, E.; Wright, R.O.; Kleinman, K.P.; Bellinger, D.; Amarasiriwardena, C.J.; Hu, H.; Rich-Edwards, J.W.; Gillman, M.W. Maternal fish consumption, hair mercury, and infant cognition in a U.S. Cohort. *Environ. Health Perspect.* **2005**, *113*, 1376–1380. [CrossRef]

68. Birch, E.E.; Hoffman, D.R.; Castaneda, Y.S.; Fawcett, S.L.; Birch, D.G.; Uauy, R.D. A randomized controlled trial of long-chain polyunsaturated fatty acid supplementation of formula in term infants after weaning at 6 wk of age. *Am. J. Clin. Nutr.* **2002**, *75*, 570–580. [CrossRef]

69. Hoffman, D.R.; Theuer, R.C.; Castaneda, Y.S.; Wheaton, D.H.; Bosworth, R.G.; O'Connor, A.R.; Morale, S.E.; Wiedemann, L.E.; Birch, E.E. Maturation of visual acuity is accelerated in breast-fed term infants fed baby food containing DHA-enriched egg yolk. *J. Nutr.* **2004**, *134*, 2307–2313. [CrossRef]

70. Birch, E.E.; Garfield, S.; Hoffman, D.R.; Uauy, R.; Birch, D.G. A randomized controlled trial of early dietary supply of long-chain polyunsaturated fatty acids and mental development in term infants. *Dev. Med. Child. Neurol.* **2000**, *42*, 174–181. [CrossRef]

71. Birch, E.E.; Garfield, S.; Castaneda, Y.; Hughbanks-Wheaton, D.; Uauy, R.; Hoffman, D. Visual acuity and cognitive outcomes at 4 years of age in a double-blind, randomized trial of long-chain polyunsaturated fatty acid-supplemented infant formula. *Early Hum. Dev.* **2007**, *83*, 279–284. [CrossRef] [PubMed]

72. Auestad, N.; Scott, D.T.; Janowsky, J.S.; Jacobsen, C.; Carroll, R.E.; Montalto, M.B.; Halter, R.; Qiu, W.; Jacobs, J.R.; Connor, W.E.; et al. Visual, cognitive, and language assessments at 39 months: A follow-up study of children fed formulas containing long-chain polyunsaturated fatty acids to 1 year of age. *Pediatrics* **2003**, *112*, e177–e183. [CrossRef] [PubMed]

73. Birch, E.E.; Carlson, S.E.; Hoffman, D.R.; Fitzgerald-Gustafson, K.M.; Fu, V.L.; Drover, J.R.; Castaneda, Y.S.; Minns, L.; Wheaton, D.K.; Mundy, D.; et al. The DIAMOND (DHA Intake And Measurement Of Neural Development) Study: A double-masked, randomized controlled clinical trial of the maturation of infant visual acuity as a function of the dietary level of docosahexaenoic acid. *Am. J. Clin. Nutr.* **2010**, *91*, 848–859. [CrossRef] [PubMed]

74. Henriksen, C.; Haugholt, K.; Lindgren, M.; Aurvag, A.K.; Ronnestad, A.; Gronn, M.; Solberg, R.; Moen, A.; Nakstad, B.; Berge, R.K.; et al. Improved cognitive development among preterm infants attributable to early supplementation of human milk with docosahexaenoic acid and arachidonic acid. *Pediatrics* **2008**, *121*, 1137–1145. [CrossRef]

75. Fang, P.C.; Kuo, H.K.; Huang, C.B.; Ko, T.Y.; Chen, C.C.; Chung, M.Y. The effect of supplementation of docosahexaenoic acid and arachidonic acid on visual acuity and neurodevelopment in larger preterm infants. *Chang. Gung Med. J.* **2005**, *28*, 708–715.

76. Meldrum, S.J.; D'Vaz, N.; Simmer, K.; Dunstan, J.A.; Hird, K.; Prescott, S.L. Effects of high-dose fish oil supplementation during early infancy on neurodevelopment and language: A randomised controlled trial. *Br. J. Nutr.* **2012**, *108*, 1443–1454. [CrossRef]

77. Isaacs, E.B.; Ross, S.; Kennedy, K.; Weaver, L.T.; Lucas, A.; Fewtrell, M.S. 10-year cognition in preterms after random assignment to fatty acid supplementation in infancy. *Pediatrics* **2011**, *128*, e890–e898. [CrossRef]

78. Faldella, G.; Govoni, M.; Alessandroni, R.; Marchiani, E.; Salvioli, G.P.; Biagi, P.L.; Spano, C. Visual evoked potentials and dietary long chain polyunsaturated fatty acids in preterm infants. *Arch. Dis. Child. Fetal Neonatal Ed.* **1996**, *75*, F108–F112. [CrossRef]

79. Lassek, W.D.; Gaulin, S.J. Linoleic and docosahexaenoic acids in human milk have opposite relationships with cognitive test performance in a sample of 28 countries. *Prostaglandins Leukot. Essent. Fat. Acids* **2014**, *91*, 195–201. [CrossRef]

80. Yamashima, T. Dual effects of the non-esterified fatty acid receptor 'GPR40' for human health. *Prog. Lipid Res.* **2015**, *58*, 40–50. [CrossRef]

81. Zamberletti, E.; Piscitelli, F.; De Castro, V.; Murru, E.; Gabaglio, M.; Colucci, P.; Fanali, C.; Prini, P.; Bisogno, T.; Maccarrone, M.; et al. Lifelong imbalanced LA/ALA intake impairs emotional and cognitive behavior via changes in brain endocannabinoid system. *J. Lipid Res.* **2017**, *58*, 301–316. [CrossRef] [PubMed]

82. Malerba, G.; Schaeffer, L.; Xumerle, L.; Klopp, N.; Trabetti, E.; Biscuola, M.; Cavallari, U.; Galavotti, R.; Martinelli, N.; Guarini, P.; et al. SNPs of the FADS gene cluster are associated with polyunsaturated fatty acids in a cohort of patients with cardiovascular disease. *Lipids* **2008**, *43*, 289–299. [CrossRef] [PubMed]

83. Caspi, A.; Williams, B.; Kim-Cohen, J.; Craig, I.W.; Milne, B.J.; Poulton, R.; Schalkwyk, L.C.; Taylor, A.; Werts, H.; Moffitt, T.E. Moderation of breastfeeding effects on the IQ by genetic variation in fatty acid metabolism. *Proc. Natl. Acad. Sci. USA* **2007**, *104*, 18860–18865. [CrossRef] [PubMed]

84. Harbild, H.L.; Harslof, L.B.; Christensen, J.H.; Kannass, K.N.; Lauritzen, L. Fish oil-supplementation from 9 to 12 months of age affects infant attention in a free-play test and is related to change in blood pressure. *Prostaglandins Leukot Essent Fatty Acids* **2013**, *89*, 327–333. [CrossRef]

85. Ryan, A.S.; Nelson, E.B. Assessing the effect of docosahexaenoic acid on cognitive functions in healthy, preschool children: A randomized, placebo-controlled, double-blind study. *Clin. Pediatrics (Phila)* **2008**, *47*, 355–362. [CrossRef]

86. Parletta, N.; Cooper, P.; Gent, D.N.; Petkov, J.; O'Dea, K. Effects of fish oil supplementation on learning and behaviour of children from Australian Indigenous remote community schools: A randomised controlled trial. *Prostaglandins Leukot. Essent. Fat. Acids* **2013**, *89*, 71–79. [CrossRef]

87. Milte, C.M.; Parletta, N.; Buckley, J.D.; Coates, A.M.; Young, R.M.; Howe, P.R. Eicosapentaenoic and docosahexaenoic acids, cognition, and behavior in children with attention-deficit/hyperactivity disorder: A randomized controlled trial. *Nutrition* **2012**, *28*, 670–677. [CrossRef]

88. Williams, P.J.; Bulmer, J.N.; Innes, B.A.; Broughton Pipkin, F. Possible roles for folic acid in the regulation of trophoblast invasion and placental development in normal early human pregnancy. *Biol. Reprod.* **2011**, *84*, 1148–1153. [CrossRef]

89. Khong, Y.; Brosens, I. Defective deep placentation. *Best Pract. Res. Clin. Obstet. Gynaecol* **2011**, *25*, 301–311. [CrossRef]

90. Mainigi, M.A.; Olalere, D.; Burd, I.; Sapienza, C.; Bartolomei, M.; Coutifaris, C. Peri-implantation hormonal milieu: Elucidating mechanisms of abnormal placentation and fetal growth. *Biol. Reprod.* **2014**, *90*, 26. [CrossRef]

91. Alfaidy, N.; Hoffmann, P.; Boufettal, H.; Samouh, N.; Aboussaouira, T.; Benharouga, M.; Feige, J.J.; Brouillet, S. The multiple roles of EG-VEGF/PROK1 in normal and pathological placental angiogenesis. *BioMed Res. Int.* **2014**, *2014*, 451906. [CrossRef] [PubMed]

92. Johnsen, G.M.; Basak, S.; Weedon-Fekjaer, M.S.; Staff, A.C.; Duttaroy, A.K. Docosahexaenoic acid stimulates tube formation in first trimester trophoblast cells, HTR8/SVneo. *Placenta* **2011**, *32*, 626–632. [CrossRef] [PubMed]

93. Innis, S.M. Essential fatty acids in growth and development. *Prog. Lipid Res.* **1991**, *30*, 39–103. [CrossRef]

94. Basak, S.; Duttaroy, A.K. Effects of fatty acids on angiogenic activity in the placental extravillious trophoblast cells. *Prostaglandins Leukot. Essent. Fat. Acids* **2013**, *88*, 155–162. [CrossRef]

95. Spencer, L.; Mann, C.; Metcalfe, M.; Webb, M.; Pollard, C.; Spencer, D.; Berry, D.; Steward, W.; Dennison, A. The effect of omega-3 FAs on tumour angiogenesis and their therapeutic potential. *Eur. J. Cancer* **2009**, *45*, 2077–2086. [CrossRef]

96. Jing, K.; Wu, T.; Lim, K. Omega-3 polyunsaturated fatty acids and cancer. *Anticancer Agents Med. Chem.* **2013**, *13*, 1162–1177. [CrossRef]

97. Calviello, G.; Di Nicuolo, F.; Gragnoli, S.; Piccioni, E.; Serini, S.; Maggiano, N.; Tringali, G.; Navarra, P.; Ranelletti, F.O.; Palozza, P. n-3 PUFAs reduce VEGF expression in human colon cancer cells modulating the COX-2/PGE2 induced ERK-1 and -2 and HIF-1alpha induction pathway. *Carcinogenesis* **2004**, *25*, 2303–2310. [CrossRef]

98. Hasan, A.U.; Ohmori, K.; Konishi, K.; Igarashi, J.; Hashimoto, T.; Kamitori, K.; Yamaguchi, F.; Tsukamoto, I.; Uyama, T.; Ishihara, Y.; et al. Eicosapentaenoic acid upregulates VEGF-A through both GPR120 and PPARgamma mediated pathways in 3T3-L1 adipocytes. *Mol. Cell. Endocrinol.* **2015**, *406*, 10–18. [CrossRef]

99. Sapieha, P.; Stahl, A.; Chen, J.; Seaward, M.R.; Willett, K.L.; Krah, N.M.; Dennison, R.J.; Connor, K.M.; Aderman, C.M.; Liclican, E.; et al. 5-Lipoxygenase metabolite 4-HDHA is a mediator of the antiangiogenic effect of omega-3 polyunsaturated fatty acids. *Sci. Transl. Med.* **2011**, *3*, 69ra12. [CrossRef]

100. Dutta-Roy, A.K. Transport mechanisms for long-chain polyunsaturated fatty acids in the human placenta. *Am. J. Clin. Nutr.* **2000**, *71*, 315S–322S. [CrossRef]

101. Campbell, F.M.; Bush, P.G.; Veerkamp, J.H.; Dutta-Roy, A.K. Detection and cellular localization of plasma membrane-associated and cytoplasmic fatty acid-binding proteins in human placenta. *Placenta* **1998**, *19*, 409–415. [CrossRef]

102. Campbell, F.M.; Dutta-Roy, A.K. Plasma membrane fatty acid-binding protein (FABPpm) is exclusively located in the maternal facing membranes of the human placenta. *FEBS Lett.* **1995**, *375*, 227–230. [CrossRef]

103. Campbell, F.M.; Gordon, M.J.; Dutta-Roy, A.K. Placental membrane fatty acid-binding protein preferentially binds arachidonic and docosahexaenoic acids. *Life Sci.* **1998**, *63*, 235–240. [CrossRef]

104. Sánchez-Campillo, M.; Ruiz-Palacios, M.; Ruiz-Alcaraz, A.J.; Prieto-Sánchez, M.T.; Blanco-Carnero, J.E.; Zornoza, M.; Ruiz-Pastor, M.J.; Demmelmair, H.; Sánchez-Solís, M.; Koletzko, B.; et al. Child Head Circumference and Placental MFSD2a Expression Are Associated to the Level of MFSD2a in Maternal Blood During Pregnancy. *Front. Endocrinol.* **2020**, *11*, 38. [CrossRef] [PubMed]

105. Prieto-Sanchez, M.T.; Ruiz-Palacios, M.; Blanco-Carnero, J.E.; Pagan, A.; Hellmuth, C.; Uhl, O.; Peissner, W.; Ruiz-Alcaraz, A.J.; Parrilla, J.J.; Koletzko, B.; et al. Placental MFSD2a transporter is related to decreased DHA in cord blood of women with treated gestational diabetes. *Clin. Nutr.* **2017**, *36*, 513–521. [CrossRef]

106. Duttaroy, A.K. Fatty acid-activated nuclear transcription factors and their roles in human placenta. *Eur. J. Lipid Sci. Technol.* **2006**, *108*, 70–83. [CrossRef]

107. Olsen, S.F.; Joensen, H.D. High liveborn birth weights in the Faroes: A comparison between birth weights in the Faroes and in Denmark. *J. Epidemiol. Community Health* **1985**, *39*, 27–32. [CrossRef]

108. Greenberg, J.A.; Bell, S.J.; Ausdal, W.V. Omega-3 Fatty Acid supplementation during pregnancy. *Rev. Obstet. Gynecol.* **2008**, *1*, 162–169.

109. Dhobale, M.V.; Wadhwani, N.; Mehendale, S.S.; Pisal, H.R.; Joshi, S.R. Reduced levels of placental long chain polyunsaturated fatty acids in preterm deliveries. *Prostaglandins Leukot. Essent. Fat. Acids* **2011**, *85*, 149–153. [CrossRef]

110. Koletzko, B.; Cetin, I.; Brenna, J.T. Dietary fat intakes for pregnant and lactating women. *Br. J. Nutr.* **2007**, *98*, 873–877. [CrossRef]

111. Carlson, S.E. Docosahexaenoic acid supplementation in pregnancy and lactation. *Am. J. Clin. Nutr.* **2009**, *89*, 678S–684S. [CrossRef] [PubMed]

112. Innis, S.M. Fatty acids and early human development. *Early Hum. Dev.* **2007**, *83*, 761–766. [CrossRef] [PubMed]

113. Ramakrishnan, U.; Stein, A.D.; Parra-Cabrera, S.; Wang, M.; Imhoff-Kunsch, B.; Juarez-Marquez, S.; Rivera, J.; Martorell, R. Effects of docosahexaenoic acid supplementation during pregnancy on gestational age and size at birth: Randomized, double-blind, placebo-controlled trial in Mexico. *Food Nutr. Bull.* **2010**, *31*, S108–S116. [CrossRef] [PubMed]

114. Innis, S.M.; Friesen, R.W. Essential n-3 fatty acids in pregnant women and early visual acuity maturation in term infants. *Am. J. Clin. Nutr.* **2008**, *87*, 548–557. [CrossRef] [PubMed]

115. Larque, E.; Gil-Sanchez, A.; Prieto-Sanchez, M.T.; Koletzko, B. Omega 3 fatty acids, gestation and pregnancy outcomes. *Br. J. Nutr.* **2012**, *107*, S77–S84. [CrossRef]

116. Mantzioris, E.; James, M.J.; Gibson, R.A.; Cleland, L.G. Dietary substitution with an alpha-linolenic acid-rich vegetable oil increases eicosapentaenoic acid concentrations in tissues. *Am. J. Clin. Nutr.* **1994**, *59*, 1304–1309. [CrossRef]

117. Imhoff-Kunsch, B.; Briggs, V.; Goldenberg, T.; Ramakrishnan, U. Effect of n-3 long-chain polyunsaturated fatty acid intake during pregnancy on maternal, infant, and child health outcomes: A systematic review. *Paediatr. Perinat. Epidemiol.* **2012**, *26*, 91–107. [CrossRef]

118. Olsen, S.F.; Secher, N.J. Low consumption of seafood in early pregnancy as a risk factor for preterm delivery: Prospective cohort study. *BMJ* **2002**, *324*, 447. [CrossRef]

119. Rogers, I.; Emmett, P.; Ness, A.; Golding, J. Maternal fish intake in late pregnancy and the frequency of low birth weight and intrauterine growth retardation in a cohort of British infants. *J. Epidemiol. Community Health* **2004**, *58*, 486–492. [CrossRef]

120. Olsen, S.F.; Sorensen, J.D.; Secher, N.J.; Hedegaard, M.; Henriksen, T.B.; Hansen, H.S.; Grant, A. Randomised controlled trial of effect of fish-oil supplementation on pregnancy duration. *Lancet* **1992**, *339*, 1003–1007. [CrossRef]

121. Helland, I.B.; Smith, L.; Blomen, B.; Saarem, K.; Saugstad, O.D.; Drevon, C.A. Effect of supplementing pregnant and lactating mothers with n-3 very-long-chain fatty acids on children's IQ and body mass index at 7 years of age. *Pediatrics* **2008**, *122*, e472–e479. [CrossRef] [PubMed]

122. Harris, W.S.; Baack, M.L. Beyond building better brains: Bridging the docosahexaenoic acid (DHA) gap of prematurity. *J. Perinatol.* **2015**, *35*, 1–7. [CrossRef] [PubMed]

123. Holman, R.T.; Johnson, S.B.; Hatch, T.F. A case of human linolenic acid deficiency involving neurological abnormalities. *Am. J. Clin. Nutr.* **1982**, *35*, 617–623. [CrossRef]

124. Bjerve, K.S.; Fischer, S.; Alme, K. Alpha-linolenic acid deficiency in man: Effect of ethyl linolenate on plasma and erythrocyte fatty acid composition and biosynthesis of prostanoids. *Am. J. Clin. Nutr.* **1987**, *46*, 570–576. [CrossRef]

125. Burdge, G.C.; Jones, A.E.; Wootton, S.A. Eicosapentaenoic and docosapentaenoic acids are the principal products of alpha-linolenic acid metabolism in young men*. *Br. J. Nutr.* **2002**, *88*, 355–363. [CrossRef] [PubMed]

126. Igarashi, M.; DeMar, J.C., Jr.; Ma, K.; Chang, L.; Bell, J.M.; Rapoport, S.I. Docosahexaenoic acid synthesis from alpha-linolenic acid by rat brain is unaffected by dietary n-3 PUFA deprivation. *J. Lipid Res.* **2007**, *48*, 1150–1158. [CrossRef]

127. Uauy, R.; Mena, P.; Valenzuela, A. Essential fatty acids as determinants of lipid requirements in infants, children and adults. *Eur. J. Clin. Nutr.* **1999**, *53*, S66–S77. [CrossRef]

128. Ciappolino, V.; Mazzocchi, A.; Botturi, A.; Turolo, S.; Delvecchio, G.; Agostoni, C.; Brambilla, P. The Role of Docosahexaenoic Acid (DHA) on Cognitive Functions in Psychiatric Disorders. *Nutrients* **2019**, *11*, 769. [CrossRef] [PubMed]

129. McNamara, R.K.; Vannest, J.J.; Valentine, C.J. Role of perinatal long-chain omega-3 fatty acids in cortical circuit maturation: Mechanisms and implications for psychopathology. *World J. Psychiatry* **2015**, *5*, 15–34. [CrossRef]

130. Guxens, M.; Mendez, M.A.; Molto-Puigmarti, C.; Julvez, J.; Garcia-Esteban, R.; Forns, J.; Ferrer, M.; Vrijheid, M.; Lopez-Sabater, M.C.; Sunyer, J. Breastfeeding, long-chain polyunsaturated fatty acids in colostrum, and infant mental development. *Pediatrics* **2011**, *128*, e880–e889. [CrossRef]

131. Bernard, J.Y.; Armand, M.; Garcia, C.; Forhan, A.; De Agostini, M.; Charles, M.A.; Heude, B.; Group, E.M.-C.C.S. The association between linoleic acid levels in colostrum and child cognition at 2 and 3 y in the EDEN cohort. *Pediatric Res.* **2015**, *77*, 829–835. [CrossRef] [PubMed]

132. Innis, S.M. Polyunsaturated fatty acids in human milk: An essential role in infant development. *Adv. Exp. Med. Biol.* **2004**, *554*, 27–43. [PubMed]

133. Simopoulos, A.P. Importance of the omega-6/omega-3 balance in health and disease: Evolutionary aspects of diet. *World Rev. Nutr. Diet.* **2011**, *102*, 10–21.

134. FAO. *Fats and Fatty Acids in Human Nutrition. Report of an Expert Consultation*; FAO Food and Nutrition Paper; FAO: Geneva, Switzerland, 2010; Volume 91, pp. 1–166.

135. Niinivirta, K.; Isolauri, E.; Laakso, P.; Linderborg, K.; Laitinen, K. Dietary counseling to improve fat quality during pregnancy alters maternal fat intake and infant essential fatty acid status. *J. Nutr.* **2011**, *141*, 1281–1285. [CrossRef] [PubMed]

136. Hautero, U.; Laakso, P.; Linderborg, K.; Niinivirta, K.; Poussa, T.; Isolauri, E.; Laitinen, K. Proportions and concentrations of serum n-3 fatty acids can be increased by dietary counseling during pregnancy. *Eur. J. Clin. Nutr.* **2013**, *67*, 1163–1168. [CrossRef] [PubMed]

137. Oken, E.; Guthrie, L.B.; Bloomingdale, A.; Platek, D.N.; Price, S.; Haines, J.; Gillman, M.W.; Olsen, S.F.; Bellinger, D.C.; Wright, R.O. A pilot randomized controlled trial to promote healthful fish consumption during pregnancy: The Food for Thought Study. *Nutr. J.* **2013**, *12*, 33. [CrossRef]

138. Emmett, R.; Akkersdyk, S.; Yeatman, H.; Meyer, B.J. Expanding awareness of docosahexaenoic acid during pregnancy. *Nutrients* **2013**, *5*, 1098–1109. [CrossRef]

139. Miles, E.A.; Noakes, P.S.; Kremmyda, L.S.; Vlachava, M.; Diaper, N.D.; Rosenlund, G.; Urwin, H.; Yaqoob, P.; Rossary, A.; Farges, M.C.; et al. The Salmon in Pregnancy Study: Study design, subject characteristics, maternal fish and marine n-3 fatty acid intake, and marine n-3 fatty acid status in maternal and umbilical cord blood. *Am. J. Clin. Nutr.* **2011**, *94*, 1986S–1992S. [CrossRef]

140. Bascunan, K.A.; Valenzuela, R.; Chamorro, R.; Valencia, A.; Barrera, C.; Puigrredon, C.; Sandoval, J.; Valenzuela, A. Polyunsaturated fatty acid composition of maternal diet and erythrocyte phospholipid status in Chilean pregnant women. *Nutrients* **2014**, *6*, 4918–4934. [CrossRef]

141. Decsi, T.; Campoy, C.; Demmelmair, H.; Szabo, E.; Marosvolgyi, T.; Escolano, M.; Marchal, G.; Krauss-Etschmann, S.; Cruz, M.; Koletzko, B. Inverse association between trans isomeric and long-chain polyunsaturated fatty acids in pregnant women and their newborns: Data from three European countries. *Ann. Nutr. Metab.* **2011**, *59*, 107–116. [CrossRef]

142. Lagarde, M.; Bernoud, N.; Brossard, N.; Lemaitre-Delaunay, D.; Thies, F.; Croset, M.; Lecerf, J. Lysophosphatidylcholine as a preferred carrier form of docosahexaenoic acid to the brain. *J. Mol. Neurosci.* **2001**, *16*, 201–204. [CrossRef]

143. Chen, C.T.; Kitson, A.P.; Hopperton, K.E.; Domenichiello, A.F.; Trepanier, M.O.; Lin, L.E.; Ermini, L.; Post, M.; Thies, F.; Bazinet, R.P. Plasma non-esterified docosahexaenoic acid is the major pool supplying the brain. *Sci. Rep.* **2015**, *5*, 15791. [CrossRef] [PubMed]

144. Demar, J.C., Jr.; Ma, K.; Chang, L.; Bell, J.M.; Rapoport, S.I. alpha-Linolenic acid does not contribute appreciably to docosahexaenoic acid within brain phospholipids of adult rats fed a diet enriched in docosahexaenoic acid. *J. Neurochem.* **2005**, *94*, 1063–1076. [CrossRef] [PubMed]

145. Chan, J.P.; Wong, B.H.; Chin, C.F.; Galam, D.L.A.; Foo, J.C.; Wong, L.C.; Ghosh, S.; Wenk, M.R.; Cazenave-Gassiot, A.; Silver, D.L. The lysolipid transporter Mfsd2a regulates lipogenesis in the developing brain. *PLoS Biol.* **2018**, *16*, e2006443. [CrossRef]

146. Kuipers, R.S.; Luxwolda, M.F.; Offringa, P.J.; Boersma, E.R.; Dijck-Brouwer, D.A.; Muskiet, F.A. Fetal intrauterine whole body linoleic, arachidonic and docosahexaenoic acid contents and accretion rates. *Prostaglandins Leukot. Essent. Fat. Acids* **2012**, *86*, 13–20. [CrossRef]

147. Talamonti, E.; Sasso, V.; To, H.; Haslam, R.P.; Napier, J.A.; Ulfhake, B.; Pernold, K.; Asadi, A.; Hessa, T.; Jacobsson, A.; et al. Impairment of DHA synthesis alters the expression of neuronal plasticity markers and the brain inflammatory status in mice. *FASEB J. Off. Publ. Fed. Am. Soc. Exp. Biol.* **2020**, *34*, 2024–2040. [CrossRef]

148. Labrousse, V.F.; Leyrolle, Q.; Amadieu, C.; Aubert, A.; Sere, A.; Coutureau, E.; Grégoire, S.; Bretillon, L.; Pallet, V.; Gressens, P.; et al. Dietary omega-3 deficiency exacerbates inflammation and reveals spatial memory deficits in mice exposed to lipopolysaccharide during gestation. *Brain Behav. Immun.* **2018**, *73*, 427–440. [CrossRef]

149. Kitajka, K.; Sinclair, A.J.; Weisinger, R.S.; Weisinger, H.S.; Mathai, M.; Jayasooriya, A.P.; Halver, J.E.; Puskás, L.G. Effects of dietary omega-3 polyunsaturated fatty acids on brain gene expression. *Proc. Natl. Acad. Sci. USA* **2004**, *101*, 10931–10936. [CrossRef]

150. Bhatia, H.S.; Agrawal, R.; Sharma, S.; Huo, Y.X.; Ying, Z.; Gomez-Pinilla, F. Omega-3 fatty acid deficiency during brain maturation reduces neuronal and behavioral plasticity in adulthood. *PLoS ONE* **2011**, *6*, e28451. [CrossRef]

151. Tian, C.; Fan, C.; Liu, X.; Xu, F.; Qi, K. Brain histological changes in young mice submitted to diets with different ratios of n-6/n-3 polyunsaturated fatty acids during maternal pregnancy and lactation. *Clin. Nutr.* **2011**, *30*, 659–667. [CrossRef]

152. Dragano, N.R.V.; Solon, C.; Ramalho, A.F.; de Moura, R.F.; Razolli, D.S.; Christiansen, E.; Azevedo, C.; Ulven, T.; Velloso, L.A. Polyunsaturated fatty acid receptors, GPR40 and GPR120, are expressed in the hypothalamus and control energy homeostasis and inflammation. *J. Neuroinflamm.* **2017**, *14*, 91. [CrossRef] [PubMed]

153. Coti Bertrand, P.; O'Kusky, J.R.; Innis, S.M. Maternal dietary (n-3) fatty acid deficiency alters neurogenesis in the embryonic rat brain. *J. Nutr.* **2006**, *136*, 1570–1575. [CrossRef]

154. Carrie, I.; Clement, M.; de Javel, D.; Frances, H.; Bourre, J.M. Specific phospholipid fatty acid composition of brain regions in mice. Effects of n-3 polyunsaturated fatty acid deficiency and phospholipid supplementation. *J. Lipid Res.* **2000**, *41*, 465–472.

155. Giltay, E.J.; Gooren, L.J.G.; Toorians, A.W.F.T.; Katan, M.B.; Zock, P.L. Docosahexaenoic acid concentrations are higher in women than in men because of estrogenic effects. *Am. J. Clin. Nutr.* **2004**, *80*, 1167–1174. [CrossRef]

156. Bondi, C.O.; Taha, A.Y.; Tock, J.L.; Totah, N.K.; Cheon, Y.; Torres, G.E.; Rapoport, S.I.; Moghaddam, B. Adolescent behavior and dopamine availability are uniquely sensitive to dietary omega-3 fatty acid deficiency. *Biol. Psychiatry* **2014**, *75*, 38–46. [CrossRef] [PubMed]

157. Auguste, S.; Sharma, S.; Fisette, A.; Fernandes, M.F.; Daneault, C.; Des Rosiers, C.; Fulton, S. Perinatal deficiency in dietary omega-3 fatty acids potentiates sucrose reward and diet-induced obesity in mice. *Int. J. Dev. Neurosci. Off. J. Int. Soc. Dev. Neurosci.* **2018**, *64*, 8–13. [CrossRef] [PubMed]

158. Harauma, A.; Moriguchi, T. Dietary n-3 fatty acid deficiency in mice enhances anxiety induced by chronic mild stress. *Lipids* **2011**, *46*, 409–416. [CrossRef]

159. Kawakita, E.; Hashimoto, M.; Shido, O. Docosahexaenoic acid promotes neurogenesis in vitro and in vivo. *Neuroscience* **2006**, *139*, 991–997. [CrossRef]

160. Moriguchi, T.; Salem, N., Jr. Recovery of brain docosahexaenoate leads to recovery of spatial task performance. *J. Neurochem.* **2003**, *87*, 297–309. [CrossRef]

161. Braarud, H.C.; Markhus, M.W.; Skotheim, S.; Stormark, K.M.; Froyland, L.; Graff, I.E.; Kjellevold, M. Maternal DHA Status during Pregnancy Has a Positive Impact on Infant Problem Solving: A Norwegian Prospective Observation Study. *Nutrients* **2018**, *10*, 529. [CrossRef]

162. Jensen, C.L.; Voigt, R.G.; Llorente, A.M.; Peters, S.U.; Prager, T.C.; Zou, Y.L.; Rozelle, J.C.; Turcich, M.R.; Fraley, J.K.; Anderson, R.E.; et al. Effects of early maternal docosahexaenoic acid intake on neuropsychological status and visual acuity at five years of age of breast-fed term infants. *J. Pediatrics* **2010**, *157*, 900–905. [CrossRef] [PubMed]

163. Ogundipe, E.; Tusor, N.; Wang, Y.; Johnson, M.R.; Edwards, A.D.; Crawford, M.A. Randomized controlled trial of brain specific fatty acid supplementation in pregnant women increases brain volumes on MRI scans of their newborn infants. *Prostaglandins Leukot. Essent. Fat. Acids* **2018**, *138*, 6–13. [CrossRef] [PubMed]

164. Colombo, J.; Shaddy, D.J.; Gustafson, K.; Gajewski, B.J.; Thodosoff, J.M.; Kerling, E.; Carlson, S.E. The Kansas University DHA Outcomes Study (KUDOS) clinical trial: Long-term behavioral follow-up of the effects of prenatal DHA supplementation. *Am. J. Clin. Nutr.* **2019**, *109*, 1380–1392. [CrossRef] [PubMed]

165. Ostadrahimi, A.; Salehi-Pourmehr, H.; Mohammad-Alizadeh-Charandabi, S.; Heidarabady, S.; Farshbaf-Khalili, A. The effect of perinatal fish oil supplementation on neurodevelopment and growth of infants: A randomized controlled trial. *Eur. J. Nutr.* **2018**, *57*, 2387–2397. [CrossRef]

166. Mulder, K.A.; Elango, R.; Innis, S.M. Fetal DHA inadequacy and the impact on child neurodevelopment: A follow-up of a randomised trial of maternal DHA supplementation in pregnancy. *Br. J. Nutr.* **2018**, *119*, 271–279. [CrossRef]

167. Smuts, C.M.; Borod, E.; Peeples, J.M.; Carlson, S.E. High-DHA eggs: Feasibility as a means to enhance circulating DHA in mother and infant. *Lipids* **2003**, *38*, 407–414. [CrossRef]

168. Judge, M.P.; Harel, O.; Lammi-Keefe, C.J. A docosahexaenoic acid-functional food during pregnancy benefits infant visual acuity at four but not six months of age. *Lipids* **2007**, *42*, 117–122. [CrossRef]

169. Judge, M.P.; Harel, O.; Lammi-Keefe, C.J. Maternal consumption of a docosahexaenoic acid-containing functional food during pregnancy: Benefit for infant performance on problem-solving but not on recognition memory tasks at age 9 mo. *Am. J. Clin. Nutr.* **2007**, *85*, 1572–1577. [CrossRef]

170. Atalah, S.E.; Araya, B.M.; Rosselot, P.G.; Araya, L.H.; Vera, A.G.; Andreu, R.R.; Barba, G.C.; Rodriguez, L. Consumption of a DHA-enriched milk drink by pregnant and lactating women, on the fatty acid composition of red blood cells, breast milk, and in the newborn. *Arch. Latinoam. Nutr.* **2009**, *59*, 271–277.

171. Echeverria, F.; Valenzuela, R.; Catalina Hernandez-Rodas, M.; Valenzuela, A. Docosahexaenoic acid (DHA), a fundamental fatty acid for the brain: New dietary sources. *Prostaglandins Leukot. Essent. Fat. Acids* **2017**, *124*, 1–10. [CrossRef]

172. Eilander, A.; Hundscheid, D.C.; Osendarp, S.J.; Transler, C.; Zock, P.L. Effects of n-3 long chain polyunsaturated fatty acid supplementation on visual and cognitive development throughout childhood: A review of human studies. *Prostaglandins Leukot. Essent. Fat. Acids* **2007**, *76*, 189–203. [CrossRef] [PubMed]

173. Jasani, B.; Simmer, K.; Patole, S.K.; Rao, S.C. Long chain polyunsaturated fatty acid supplementation in infants born at term. *Cochrane Database Syst. Rev.* **2017**, *3*, CD000376. [CrossRef] [PubMed]

174. Khandelwal, S.; Swamy, M.K.; Patil, K.; Kondal, D.; Chaudhry, M.; Gupta, R.; Divan, G.; Kamate, M.; Ramakrishnan, L.; Bellad, M.B.; et al. The impact of DocosaHexaenoic Acid supplementation during pregnancy and lactation on Neurodevelopment of the offspring in India (DHANI): Trial protocol. *BMC Pediatrics* **2018**, *18*, 261. [CrossRef]

175. Khandelwal, S.; Kondal, D.; Chaudhry, M.; Patil, K.; Swamy, M.K.; Metgud, D.; Jogalekar, S.; Kamate, M.; Divan, G.; Gupta, R.; et al. Effect of Maternal Docosahexaenoic Acid (DHA) Supplementation on Offspring Neurodevelopment at 12 Months in India: A Randomized Controlled Trial. *Nutrients* **2020**, *12*, 3041. [CrossRef]

176. Clandinin, M.T.; Van Aerde, J.E.; Merkel, K.L.; Harris, C.L.; Springer, M.A.; Hansen, J.W.; Diersen-Schade, D.A. Growth and development of preterm infants fed infant formulas containing docosahexaenoic acid and arachidonic acid. *J. Pediatrics* **2005**, *146*, 461–468. [CrossRef] [PubMed]

177. Baack, M.L.; Puumala, S.E.; Messier, S.E.; Pritchett, D.K.; Harris, W.S. Daily Enteral DHA Supplementation Alleviates Deficiency in Premature Infants. *Lipids* **2016**, *51*, 423–433. [CrossRef]

Publisher's Note: MDPI stays neutral with regard to jurisdictional claims in published maps and institutional affiliations.

Review

Maternal Supply of Both Arachidonic and Docosahexaenoic Acids Is Required for Optimal Neurodevelopment

Sanjay Basak, Rahul Mallick, Antara Banerjee, Surajit Pathak and Asim K. Duttaroy

1 Molecular Biology Division, ICMR-National Institute of Nutrition, Indian Council of Medical Research, Hyderabad 500 007, India; sba_bioc@yahoo.com
2 A.I. Virtanen Institute for Molecular Sciences, University of Eastern Finland, 70210 Kuopio, Finland; rahul.mallick@uef.fi
3 Department of Medical Biotechnology, Faculty of Allied Health Sciences, Chettinad Academy of Research and Education (CARE), Chettinad Hospital and Research Institute (CHRI), Kelambakkam, Chennai 603 103, India; antara.banerjee27@gmail.com (A.B.); surajit.pathak@gmail.com (S.P.)
4 Department of Nutrition, Institute of Basic Medical Sciences, Faculty of Medicine, University of Oslo, 0317 Oslo, Norway
* Correspondence: a.k.duttaroy@medisin.uio.no; Tel.: +47-22-82-15-47

Abstract: During the last trimester of gestation and for the first 18 months after birth, both docosahexaenoic acid,22:6n-3 (DHA) and arachidonic acid,20:4n-6 (ARA) are preferentially deposited within the cerebral cortex at a rapid rate. Although the structural and functional roles of DHA in brain development are well investigated, similar roles of ARA are not well documented. The mode of action of these two fatty acids and their derivatives at different structural–functional roles and their levels in the gene expression and signaling pathways of the brain have been continuously emanating. In addition to DHA, the importance of ARA has been much discussed in recent years for fetal and postnatal brain development and the maternal supply of ARA and DHA. These fatty acids are also involved in various brain developmental processes; however, their mechanistic cross talks are not clearly known yet. This review describes the importance of ARA, in addition to DHA, in supporting the optimal brain development and growth and functional roles in the brain.

Keywords: arachidonic acid,20:4n-6; brain; docosahexaenoic acid,22:6n-3; fetus; maternal diet; cognitive; infants; neurodevelopment; neurogenesis

Citation: Basak, S.; Mallick, R.; Banerjee, A.; Pathak, S.; Duttaroy, A.K. Maternal Supply of Both Arachidonic and Docosahexaenoic Acids Is Required for Optimal Neurodevelopment. *Nutrients* **2021**, *13*, 2061. https://doi.org/10.3390/nu13062061

Academic Editor: Andrew J. Sinclair

Received: 25 April 2021
Accepted: 14 June 2021
Published: 16 June 2021

Publisher's Note: MDPI stays neutral with regard to jurisdictional claims in published maps and institutional affiliations.

1. Introduction

The neurodevelopmental process involves a complex interplay between nutrients, genes, and environmental factors that result in the optimal growth, development, and maturation of the brain. The development of the brain in utero critically depends on the maternal supply of several components for its well-regulated structural–developmental process, characterized by specifically defined developmental periods, growth, a cellular signaling system, and maturation. During the brain growth spurt, neurodevelopment is particularly vulnerable to nutritional deficiencies [1].

The long-chain polyunsaturated fatty acids (LCPUFAs), docosahexaenoic acid,22:6n-3 (DHA), and arachidonic acid,20:4n-6 (ARA) are important nutrients required for fetal brain growth and development. The accumulation of DHA and ARA in the fetal brain predominantly occurs in the third trimester of a human pregnancy. The de novo synthesis of these LCPUFAs seems low in a growing fetus and placenta [2]; the maternal intake of these fatty acids contributes a significant share for brain development. Maternal DHA and ARA are accumulated rapidly within the cerebral cortex during the last trimester of pregnancy and postnatal 18 months [2–4]. The dietary balance of DHA and ARA intake and their interactions are thought to be important for the development and function of the brain. Several experimental studies suggested a crucial involvement of these two fatty acids in neural membrane formation and various roles of their metabolites, production of eicosanoids, and their influence on depression- and anxiety-related behaviors. Moreover, multiple trials have found that higher plasma or erythrocyte DHA levels positively correlate with infant neurocognitive outcomes [5–10].

High levels of DHA in the brain are achieved during early life and are maintained throughout life. DHA accretion to the brain continues into childhood, and the incorporation of DHA is still high despite its reduced accumulation rate. The preferential transfer of maternal DHA and ARA by the placenta to fetal circulation and its mechanisms are reviewed extensively [11,12]. LCPUFA content in human milk also regulate the amount of DHA and ARA transferred to the infant during breastfeeding [13]. However, this phenomenon may depend on an optimal maternal dietary intake of DHA from the supplement or marine fish [14], whereas ARA levels are usually maintained due to high n-6/n-3 ratios in the diet.

In the central nervous system (CNS), the proportion of DHA with other membrane fatty acids increases as the brain size increases. The increase in the proportion of these fatty acids continues for the second year of life. DHA has significant neurobiochemical roles in ion channel and receptor functions, the release of neurotransmitters, synaptic plasticity, and gene expression in the neurons. Both DHA and ARA of synaptic membrane phospholipids are released as free fatty acids (FFAs) by activated phospholipase A_2 (PLA_2) and are converted to different bioactive metabolites. These liberated fatty acids and the metabolites play critical roles during ischemia seizure activity, inflammation, and other types of brain disorders. Figure 1 describes the putative roles of DHA in neural membranes.

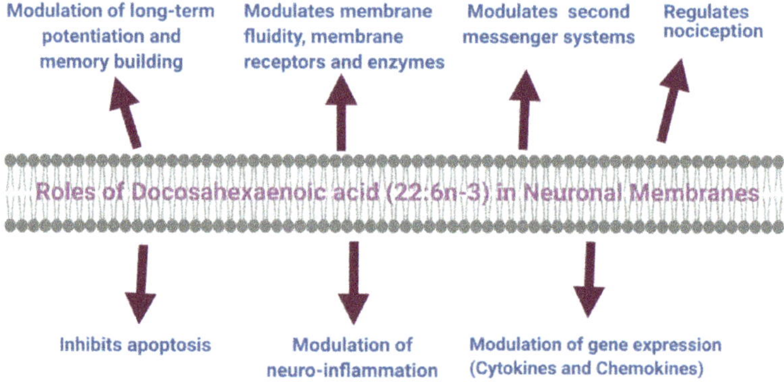

Figure 1. DHA modulates several aspects of structural and functional activities of neuronal membrane.

Maternal n-3 PUFA deficiency during pregnancy was associated with impaired brain development in offspring [15] and defective neuroblast migration [16]. The deficiency of n-3 LCPUFAs during development causes hypomyelination in the brain, resulting in mood and anxiety disorders [17,18]. Consuming a diet containing a high amount of n-3 fatty acids during pregnancy protected infants against the detrimental effects of maternal stress [19].

The essentiality of DHA is well recognized in childhood and adult life, as its deficiency causes cognitive decline and other psychiatric disorders [20]. Plasma DHA levels are inversely correlated with depressive symptoms in infants and adolescents with bipolar disorder [21]. Human breast milk containing higher n-3 and n-6 LCPUFAs was associated with decreased infant despair and distress [22]. The impact of n-3 PUFAs in human milk in influencing infant mood or anxiety is still not clearly established, since cortisol levels in milk are also associated with infant temperament [23]. Moreover, the variation in the fatty acid composition of mothers' milk may play an essential role in the outcome of offspring's mental status and overall health.

Although n-3 LCPUFAs supplementation is beneficial in preventing and treating major depression, bipolar disorder, and anxiety disorders in adults [24,25], much less is known about how the imbalance of these LCPUFA levels impact the mood and behavior of infants. Studies in experimental models suggested that early exposure to n-3 fatty acids had a lasting effect on temperament and behavioral phenotypes of offspring [26]. Interventional studies in adults also showed an association between the n-3 PUFA status with improved mood and mental health [17,26].

Though precise molecular mechanisms are not well defined, DHA and its bioactive derivatives play various essential structural and functional roles in the brain [12,27,28]. High DHA levels in the phospholipids of synaptic membranes provide membrane flexibility and improve the efficiency of protein–protein interaction necessary for signal transduction [27]. Different aspects of metabolism, and the structure–function of the brain depend on optimum ARA and DHA levels and interactions between their metabolites [29]. Cognitive benefits from supplementation of combined ARA and DHA in early life have also been observed through early and middle childhood [30,31]. Furthermore, several human brain diseases, such as Alzheimer's and bipolar disorders, involve disturbed n-3 and n-6 LCPUFA uptake and metabolism. Therefore, understanding the dynamics of the maternal supply of DHA and ARA to the developing brain may be important for managing brain disorders. An additional neuroprotective mechanism may involve bioactive molecules derived from DHA and ARA, which are also involved in several cellular neuronal biochemical processes. These bioactive derivatives modify the functions of several genes in the brain by acting as ligands for transcription factors involved in critical brain functions, including signal transduction and synaptic plasticity.

A literature search was performed on the PubMed database by using search terms such as DHA, ARA, brain development, infant, fetal, breastfeeding nutrients deficiency, DHA and ARA supplementation. All types of articles related to human and mechanistic studies on models were included for evaluation. The articles for which full text was not available or not reported in English were excluded. The articles retrieved in the first round of searches identified additional references by a manual search among the cited references. This review describes the latest development of the interplay of DHA and ARA transfer and their impacts on brain development, their complementarity, the structure–function relationship, and their mechanisms of action in the brain. Moreover, the evidence of the essentiality of ARA in brain development is summarized.

2. Maternal Delivery of DHA and ARA to the Developing Brain

Both DHA and ARA are major components of the brain. DHA comprises 10–20% of the total fatty acid composition in the brain, whereas 9% is present in the form of ARA [32]. During the third trimester of pregnancy, these LCPUFAs preferentially accumulate in the fetus and reach higher fetal/neonatal blood levels than those in the mother [33]. Both DHA and ARA are deposited in large amounts relative to the accretion rates of other fatty acids in the fetal brain during a maximum brain growth spurt, which occurs from the last trimester in utero and continues through the breastfed postnatal life [4,34–36]. The development of the brain is critically dependent on the adequate maternal supply of LCPUFAs in this period, since their synthesis from the parent EFAs is insufficient to meet the high requirement [4,37]. The lower fetal status of both DHA and ARA is associated with neurological development [38]. Consequently, ample fetal and neonatal LCPUFA supply via transplacental transport and human milk is critically important, implying that maternal LCPUFA status should be adequate.

Maternal dietary intakes of fatty acids influence the fatty acid composition of breast milk and plasma levels of lactating women and their infants [39]. DHA incorporation in the neuronal membrane in early fetal life solely depends on placental transfer [11], breastfeeding, and the endogenous synthesis of DHA [40–42]. However, DHA accretion in the CNS depends on the dietary provision of DHA, i.e., on the duration and concentration of DHA supplementation. A similar nutritional dependency is absent for ARA accretion in the brain, i.e., the dosage and duration of postnatal ARA supplementation do not affect ARA accretion in the CNS [43].

A robust linear relationship between maternal DHA level and umbilical cord blood phospholipid was reported [44]. Both ARA and DHA also have a role in the early placentation process, in addition to their roles in fetal neurodevelopment and in the postnatal lactation period [45–47]. The high-affinity placental plasma membrane, fatty acid-binding protein (p-FABPpm), is involved in the preferential supply of both maternal DHA and ARA to the fetus [12,41,48]. Since the rapid deposition of these LCPUFAs into the brain occurs during the last trimester of pregnancy and subsequently in lactation, the maternal status must be maintained well during the critical time of brain development. The dietary intake and maternal stores of DHA are the determinants of infant blood DHA concentrations at birth [49]. Blood LCPUFAs in breast-fed infants remain higher than those in maternal circulation postnatally [50,51]. The blood levels of PUFAs of infants are higher than those in the maternal circulation postnatally during the breastfeeding period. The preferential postnatal deposition of LCPUFAs in the infant's brain is mediated via breastmilk. The dietary supplementation of DHA to pregnant and nursing mothers dose-dependently increases the DHA level in breast milk, which causes higher tissue accretion of DHA in breastfed infants with improved outcomes of mental performance [52–54]. The erythrocyte DHA status of breastfed infants is correlated with the maternal DHA status of erythrocytes during lactation; however, no such association was observed for ARA [13]. Human breast milk levels of DHA and ARA are relatively stable throughout the lactation [55]. The worldwide mean concentration of DHA is $0.3 \pm 0.2\%$, and that of ARA is $0.5 \pm 0.1\%$ in breast milk [56]. Typically, human breast milk has 1.5- to 2-fold more ARA than DHA, though the breast milk's DHA content can be higher than ARA in populations with high marine fish consumption. It has been shown that the DHA level in breast milk is directly related to the DHA content of the maternal diet [57]. Still, it is unknown whether metabolic or dietary mechanisms explain the lower variability in breast milk ARA. Breast milk ARA was not affected in lactating mothers allocated to consume increasing doses of DHA [57]. Two recent studies confirmed the different regulation of ARA and DHA in breast milk, indicating that ARA is affected by the genetic pattern in the *FADS*-gene cluster [58] and is less sensitive than DHA to dietary supplementation [59]. The essentiality of ARA in infant nutrition is supported by the observation of the potentially adverse effects in preterm infants of consuming a marine-oil-containing formula [60]. Dietary ARA has roles in growth, neuronal development, and cognitive function in infants. Both ARA and DHA are necessary for fetal development, and a deficiency in one may compromise growth [61]. It has been shown that growth deficiency induced by fish oil supplementation is related to a reduced ARA availability due to the excess of DHA [62]. Western diets usually have a much higher ratio of n-6/n-3 fatty acids [63]. Therefore, it has been proposed that supplementation during gestation should be based on intake of n-3 fatty acids.

Children's intelligence quotient (IQ) was increased by 0.8 to 1.8 points when their mothers consumed DHA from pregnancy to the lactation period and beyond [64,65]. The mean levels of DHA and ARA in breast milk are found at 0.37% and 0.55% of total fatty acids across the globe, respectively. Prospective observational studies suggested that breastfed infants had a significant neurocognitive advantage compared with formula-fed infants [5,66,67], possibly due to the higher incorporation of DHA and ARA in breast milk relative to formula milk. The association between breastfeeding and child IQ concerning FADS2 genetic profile, specifically in SNP rs174575b, was observed [68]. Breastfed infants with rs174575 C-dominant carriers achieved higher scores on standardized IQ tests than non-breastfed C-carrier infants. However, observational data are confounded by the heterogeneous composition of breast milk, and environmental factors that influence the infant's mental development.

Based on established guidelines, it is emphasized that maternal dietary DHA requirements should be increased during pregnancy and lactation. Precisely, a minimum of 200 mg of DHA per day is recommended during these periods [69]. In both full-term and preterm populations, the evidence is compelling that breastfeeding is vital for an infant's neurodevelopment.

3. The Fatty Acid Uptake System of the Brain

The de novo synthesis of DHA in the brain is almost non-existent, and therefore, it must be imported from the circulation across the blood–brain barrier (BBB) [70]. The BBB comprises endothelial cells in the capillary connected with tight junctions, astrocytic end-foot processes, pericytes, and neurons [71]. The endothelial cells offer highly selective permeability in the BBB by their specialized tight junction function. Thus, the BBB is provided with a neuronal system immune-privileged environment permitting only small molecules into the brain.

Free fatty acids (FFAs) are transported into the cytosol via cell membrane- and cytoplasmic fatty acid-binding proteins (FABPs) [12]. There are four classes of membrane fatty acid transport proteins present in the brain, such as fatty acid translocase (FAT/CD36), plasma membrane fatty acid-binding proteins (FABPpm), fatty acid transport proteins (FATPs), and several cytosolic fatty acid-binding proteins (FABPs) [41,72,73]. Additionally, another transporter, Mfsd2a (major facilitator superfamily domain-containing protein 2A), is present in the BBB [12]. Mfsd2a transports lysophosphatidylcholine (LPC)-DHA, but not unesterified DHA [12,74–76]. The DHA-LPC produced from DHA-containing phosphatidylcholine when acted upon by PLA1. A sterol regulatory element-binding protein regulates the activity of Mfsd2a to maintain a balance between de novo lipogenesis and exogenous uptake of LPC-DHA. The brain of the Mfsd2a-deficient mice had significantly reduced DHA levels and experienced loss of neurons in the hippocampus and cerebellum. The mice later developed microcephaly with severe cognitive deficits and anxiety. Altered plasma levels of Mfsd2a during pregnancy influence the placental transport of DHA and neurodevelopment in utero [77]. The maternal blood levels of Mfsd2a in the third trimester were inversely correlated to DHA-LPC in maternal plasma [77]. DHA or DHA-LPC is taken up by the endothelial cells via FAT/CD36, FATPs, or Mfsd2a. Plasma albumin binds both unesterified DHA and DHA-LPC.

Like LPC-DHA, the accretion of ARA to the brain from ARA-containing lysophospholipids is six-fold more efficient than for unesterified ARA [78]. The brain relies on different transport systems for shuttling the varied forms of the metabolized DHA across the BBB. Active pathways of DHA and ARA incorporation into the brain include members of the FABP family. In human brain microvessel endothelial cells, FAT/CD36 and FATP-1, FATP4 also transport fatty acids across the monolayer [73]. However, DHA incorporation into brain phospholipids was not affected in CD36$_{-/-}$ mice, indicating this transporter may not be involved in the transport of DHA.

FABPs involve in the FFA uptake and transport to various intracellular compartments of a cell [73,79]. The expression of various FABP genes occurs at different developmental stages of the brain, as shown in animal studies. FABP3 is expressed in the brain after birth, and its expression levels increases until adulthood [80]. FABP4 is mainly expressed in grade IV astrocytomas and normal brain tissue [81], whereas FABP5 is expressed in the mid-term embryonic rat brain and reaches its peak at birth, then gradually decreases in postnatal life [80]. FABP7 is expressed mainly in radial glial cells at the early stages of brain development [80,82]. The level of FABP7 expression decreases starkly in the neonate and adult brain [80]. FABP7 plays a role in the establishment of the radial glial fiber system [82]. Moreover, FABP7 has been suggested to be associated with the decreased survival of glioblastoma patients [83–85].

FABPs have broad binding specificity, including the binding affinity for long-chain fatty acids ($\geq$c16), eicosanoids, bile salts, and PPARs [41,86]. Both FABP3 and FABP4 bind ARA with high affinity [87,88], while FABP5 preferentially binds long-chain saturated fatty acids (c $\geq$ 16) [89]. FABP7 binds both DHA and ARA but with four-fold more affinity for DHA [90], indicating that DHA may be a preferred ligand for FABP7. Although FABP7, FABP5, and FABP3 can also bind different types of fatty acids [89,91].

FABP3, FABP5, and FABP7 are involved in both developing and mature adult brains [86,92]. Functional studies have demonstrated a variety of roles for FABPs in brain development, including the generation of neuronal and/or glial cells, differentiation, neuronal cell migration, and axis patterning. Like DHA, FABP7 increases the proliferation of neural stem cells and neural progenitor cells and differentiates into mature neurons both in vitro and in vivo [93,94].

FABPs are multi-functional proteins, and complex signaling networks and transcription factors regulate their expression. FABPs are major downstream effectors of the Reelin-Dab1/Notch pathway that involves neuron–glia crosstalk during brain development. Since LCPUFAs and several FABPs are involved in brain development and function, it is important to further elucidate their roles in brain disorders' pathogenesis.

As FABPs are involved in developing, establishing, and maintaining the central nervous system, FABPs are implicated in the pathogenesis of Down syndrome. FABP7 is overexpressed in Down syndrome adult [95] and fetal brains [96], whereas FABP3 is significantly decreased in Down syndrome adult brains [95]. Furthermore, FABP7 upregulation correlates with PKNOX1 gene-dosage imbalance in the brains of Down syndrome patients. PKNOX1 is a POU domain protein that may directly control FABP7 expression by interacting with the Pbx/POU binding element of the FABP7 promoter [96]. Human FABP7 mRNA levels were significantly upregulated in the postmortem brains of schizophrenia patients. A correlation between an SNP variant within the second exon of human FABP7 and schizophrenia pathology was observed [97]. The impairment of prepulse inhibition (PPI) occurs in many brain disorders such as Alzheimer's disease, autism, bipolar disorders, Tourette syndrome, and schizophrenia [98]. The association between FABP7 and PPI status suggests roles of the FABP7 gene in the pathology of PPI-mediated neuropsychiatric and/or neurodegenerative diseases. The association of FABP3 and FABP5 with several neurodegenerative diseases is also reported [99,100]. DHA influences brain functions and protects from different brain tumors such as astrocytoma and glioma by binding to the FABP, resulting in the activation of transcription factor PPARγ to the nucleus-reduced cell migration, growth arrest, and apoptosis of tumor cells [101]. However, the roles of these proteins in human brain development are not well known.

4. Structural and Functional Roles of DHA in the Human Brain

The brain is known as the body's fattiest organ, containing phospholipids around 2/3 of its weight. The brain harness 20% of its total energy from β-oxidation of fatty acids in the mitochondria of astrocytes. The presence of DHA on the membrane influences neuronal information transmission, signal transduction velocity, and interaction with ion channels or receptor proteins and their activity [102]. DHA is also predominantly present in cortical gray matter, representing approximately 15% of total fatty acids in the adult human prefrontal cortex. As a crucial structural ingredient of the brain, DHA comprises the regions of the cerebral cortex and synaptic membrane. Neuronal membranes have approximately 50% DHA [44]. DHA is also vital for hippocampal and cortical neurogenesis, neuronal migration, and outgrowth [93,103,104].

The brain fatty acid levels, mostly LCPUFAs, are maintained via different mechanisms, as described above. The circulating plasma levels of DHA is positively related to cognitive abilities during aging and is inversely associated with decline in cognitive function. DHA, being a part of the cell membrane phospholipid, contributes to maintaining optimal fluidity and lipid raft assembly in the membranes, membrane electrical and antigenic signals of the cells. DHA also halts cell death by stimulating cell-cycle exit in neuro-progenitor cells [93,105]. DHA is involved in monoaminergic and cholinergic systems during brain development processes [52,106,107]. DHA has a long-term effect on serotonergic and dopaminergic systems during the fetal brain development in utero [35,107,108]. Data emphasized the importance of examining the long-term critical impact on brain development due to inadequate DHA supplies to the fetus during pregnancy. DHA stimulates neurite outgrowth in cell culture systems. Neurite outgrowth is an important process in the developing nervous system and also in the regeneration of nerves. The alpha linoleneic acid, 18:3n-3 (the precursor of DHA)-restricted diet decreased neurogenesis in rat dams' fetal brains, possibly due to the deficiency of DHA [103]. DHA influences gene expression, neurotransmission and protects the brain from oxidative stress during development [109]. DHA is an essential factor for neurogenesis, phospholipid synthesis, and turnover [93,110,111].

Again, DHA can act as a ligand for peroxisome proliferator-activated receptor-gamma (PPARγ) and retinoid X receptor (RXR). RXR plays a vital role in embryonic neurogenesis, neuronal plasticity, and catecholaminergic neuron differentiation along with retinoic acid receptors. RXR is highly expressed in the hippocampus [112,113]. The PPARγ-RXR heterodimer modulates early brain development by regulating transcription genes [112,114].

DHA also protects the developing brain from peroxidative damage of lipids and proteins [115–117]. DHA and eicosapentaenoic acid,20:5n-3 (EPA) were reported as suppressors of angiogenesis in cancer cells, but they stimulate angiogenesis in the placenta [45]. However, DHA and DHA-LPC may act as pro-angiogenic and anti-angiogenic depending on the concentration and microenvironments [118].

Several epidemiological data show an inverse association of low habitual dietary intake of DHA and a higher risk of brain diseases [2,119,120]. A diet containing high amounts of n-3 fats and/or a lower amount of n-6 fats was strongly associated with the lower incidence of Alzheimer's disease and other brain diseases [121–124]. Intake of DHA improves attention deficit hyperactivity disorder (ADHD), bipolar disorder, schizophrenia, impulsive behavior, and other brain disorders [20,123,125].

Nevertheless, the data of intervention studies with DHA supplements are conflicting, despite the fact that many such studies demonstrated an apparent benefit of DHA intake in brain function. Several studies failed to reproducibly show that the absence of DHA and its metabolites are involved in various adult brain diseases. More well-designed clinical trials considering background diets and genetic makeup are needed for definitive conclusions.

5. Roles of DHA and Its Metabolites in the Brain

DHA and its metabolites play vital roles in the functional brain development of the fetus in utero and infants and healthy brain function in adults. DHA and its metabolites play significant roles in cellular and biological functions. The oxidation of DHA by lipoxygenases produces several types of metabolites such as oxylipins that regulate various biochemical processes of the brain [126].

DHA stimulates membrane-associated G-protein-coupled receptor (GPR) 120 mediated gene activation to promote anti-inflammatory activities [127,128]. DHA also activates PPARs and upregulates the expression of genes responsible for increasing insulin sensitivity and reducing plasma triglyceride level and inflammation. [44]. DHA and its metabolites' signaling pathways are involved in neurogenesis, anti-nociceptive effects, anti-apoptotic effects, the plasticity of the synapse, Ca^{2+} homeostasis in the brain, and nigrostriatal activities [129]. DHA itself and its metabolites have a broad spectrum of actions at different levels and sites in the brain [129,130]. Figure 2 shows the DHA and EPA metabolites and their function in the brain.

DHA is converted to maresin 1(MaR1), neuroprotectin D1(NPD1), and resolvins by human 12-LOX, 15-LOX, and CYP or enzymatically by aspirin-treated COX-2. They play various functions in the brain. DHA is metabolized by the P450 system, cyclooxygenase, and lipoxygenase enzymes under different metabolic conditions. 14-HDHA: 14-hydroxy-docosahexaenoic acid; 17-HDHA: 17-hydroxy-docosahexaenoic acid; 18-HEPE: 18-hydroxy-eicosapentaenoic acid; LG R6: G protein-coupled receptor 6; ALX/Fpr2: N-formyl peptide receptor 2; BLT1: leukotriene B4 receptor; COX-2: cyclooxygenases; DHA: docosahexaenoic acid; EPA: eicosapentaenoic acid; GPR32/37: G protein-coupled receptor 32/37; LC-PUFAs: long-chain polyunsaturated fatty acids; LOX: lipoxygenases.

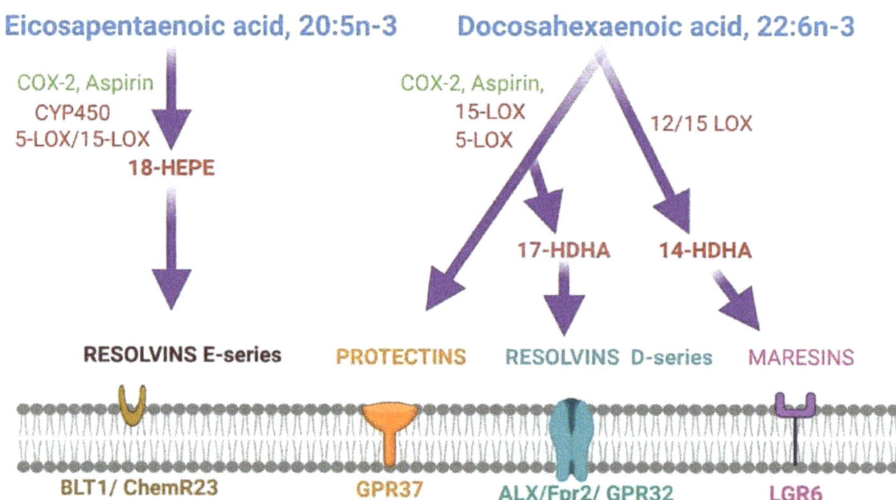

Figure 2. Metabolites of EPA and DHA, and their membrane receptors.

DHA-derived specialized pro-resolving mediators (SPMs) such as DHA epoxides, oxo-derivatives (EFOX) of DHA, neuroprostanes, ethanolamines, acylglycerols, docosahexaenoyl amides of amino acids, and branched DHA esters of hydroxy fatty acids play important roles in brain functions [131,132]. Additionally, epoxydocosapentaenoic acids (EDPs) and 22-hydroxydocosahexaenoic acids (22-HDoHEs) are produced from DHA [133,134]. DHA is mainly metabolized by enzymes such as 5-, 12- and 15-lipoxygenases (LOX), COX-2, and cytochrome P450 (CYP). As the most demanding by-products of DHA, resolvins are formed by either LOX15 or CYP or aspirin-treated COX-2 activity [28]. The LOX15-derived resolvins are homologous to CYP-, or aspirin-treated COX-2-derived resolvins [28]. As the inflammation resolution mediators, resolvins act via different G-protein coupled receptors (GPRs) [28,135]. Both resolvin D1 and aspirin-triggered resolvin D1 improve brain functions and impede neuronal death by down-regulating several factors such as NFkB, TLR4, CD200, and IL6R [136,137]. They even induce remote functional recovery after brain damage [136]. Both resolvin D2 and aspirin-triggered resolvin D2 protect from cerebral ischemic injury via phosphorylation ERK1/2. Subsequently, this pathway stimulates nNOS or eNOS to inhibit neuronal cell death and maintain BBB integrity by increasing zonula occludens-1 [138]. Resolvin D3, resolvin D5, aspirin-triggered COX-2 -derived resolvin D3, and aspirin-triggered resolvin D5 halt the neuroinflammation [139,140]. However, the functions of other resolvins are still a mystery. Maresin (MaR), the anti-inflammatory pro-resolving mediator, is produced from DHA in macrophages during the inflammation, healing, and regeneration process [141–143]. MaR1 is predominant among other maresins. MaR1 decreases LTB_4 synthesis and stimulates phagocytosis at the site of inflammation [144–146]. MaR1 enhances tissue repair by stimulating stem cell differentiation and plays an analgesic role through TRPV1-mediated response blockage [145,147]. MaR1 involves neurocognitive functions by regulating the infiltration of macrophages, regulating NF-κB signaling, oxidative stress, and cytokine release. Maresin 1 reduces neuroinflammation perioperative neurodegenerative disorders in an animal model [148]. MaR1 significantly affects the post-spinal cord injury model [149,150].

Another SPM, neuroprotectin D1(NPD1), derived from DHA, improves cell survival and cell repair in brain disorders [151]. Like MaR1, NPD1 also possesses anti-inflammatory and neuroprotective activities [152]. In response to neuroinflammation, NPD1 is produced from endogenous DHA in the retina and brain [153,154]. Besides antiviral protection, NPD1 helps in neurocognitive functions [155–157]. NDP1 blocks the progression of Alzheimer's disease by stimulating the expression of PPARγ, amyloid precursor protein-α, and reducing the β-amyloid precursor protein [156]. Figure 3 describes DHA metabolites and their global effects on gene expression and second messenger systems affecting multiple cellular functions in the brain.

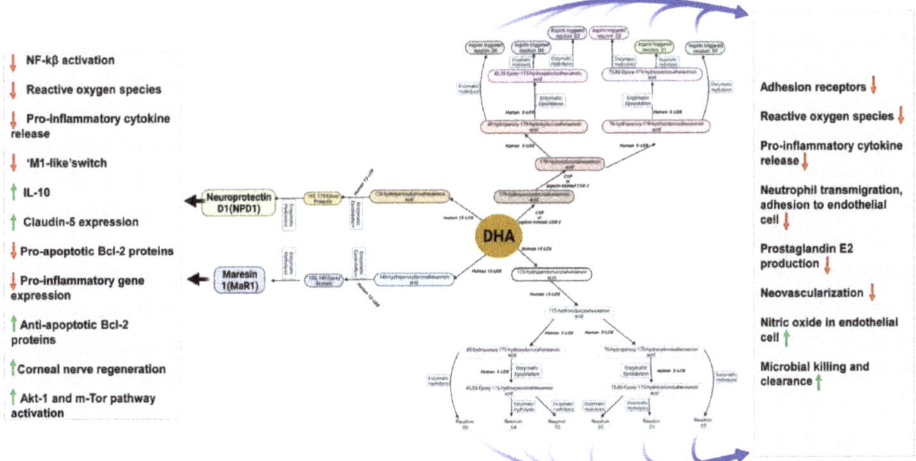

Figure 3. DHA metabolites formation and function in the brain.

Maresin 1(MaR1), neuroprotectin D1(NPD1), and resolvins are produced from DHA by human 12-LOX, 15-LOX, and CYP or aspirin-treated COX-2 enzymatically. These metabolites have multiple functions in the brain, which have been mentioned in the boxes. DHA is metabolized by the P450 system, cyclooxygenase, and lipoxygenase enzymes under different metabolic conditions.

Anti-inflammatory activities of EFOX and neuroprostanes protect neuroinflammation in various diseases, such as Parkinson's disease and Alzheimer's disease [158–160]. Another DHA derivative, docosahexaenoyl ethanolamide, improves mood, pain, inflammation status, hunger, and glucose uptake by the brain endocannabinoid system [161–166]. The function of DHA metabolites is summarized in Table 1.

DHA glyceryl ester regulates the intake of food and neuroinflammation, similarly to the way docosahexaenoyl ethanolamide uses the endocannabinoid system [167,168]. The endocannabinoid system plays an integral part in memory, cognition, and pain perception [169,170]. DHA conjugates via cannabinoid receptors reduce neuroinflammation and improve neurogenesis [171,172].

DHA is involved in alleviating short-term stress, preventing anxiety and stress in later life [132,173]. DHA is reported to improve various psychiatric disorders such as schizophrenia, mood and anxiety disorders, obsessive-compulsive disorder, ADHD, autism, aggression, hostility and impulsivity, borderline personality disorder, substance abuse, and anorexia nervosa [174]. There is strong evidence that the consumption of marine fish reduces depression [175].

The mild symptoms of ADHD are corrected with DHA supplementation [176]. DHA was also shown to improve depressive symptoms of bipolar disorder by increasing N-acetyl-aspartate brain levels without affecting mania [174]. Even IQ outcomes in children and cognitive function in the aging brain are improved by DHA supplementation [2,24]. DHA-derived anti-inflammatory eicosanoids' neuroprotection prevent Alzheimer's disease pathogenesis [177]. DHA may modulate the metabolism of cholesterol and apolipoprotein E, lipid raft assembly, and the cell signaling system in Alzheimer's disease [178,179]. DHA protects neuronal brain function by reducing NO production, calcium influx, and apoptosis while activating antioxidant enzymes such as glutathione peroxidase and glutathione reductase [180]. The second most prevalent neurodegenerative disease, Parkinson's disease, can be halted by DHA's neuroprotective role [180].

Table 1. Functions of DHA and ARA metabolites.

Metabolites	Name	Biological Effects
DHA Metabolites	Maresins	Resolution of inflammation, wound healing, analgesic effects
	Protectins	Resolution of inflammation, neuroprotection
	Resolvins	Resolution of inflammation and wound healing
	Electrophilic oxo-derivatives (EFOX) of DHA	Anti-inflammatory, anti-proliferative effects
	Epoxides	Anti-hypertensive, analgesic actions
	Neuroprostanes	Cardio-protection, wound healing
DHA conjugates	Ethanolamines and glycerol esters	Neural development, immunomodulation, metabolic effects
	Branched fatty acid esters of hydroxy fatty acids (FAHFA)	Immuno-modulation, resolution of inflammation
	N-acyl amides	Metabolic regulation, neuroprotection, neurotransmission
ARA metabolites	Lipoxins A4	Lowers neuroinflammation by inhibiting microglial activation
	Lipoxins B4	Promotes neuroprotection from acute and chronic injuries

6. DHA Deficiency in Utero and Human Brain Function

DHA deficiency is linked with different brain disorders such as major depressive and bipolar disorder [181,182]. DHA levels are positively correlated with improved learning and memory and reduced neuronal loss [24]. DHA deficiency affects epigenetic development in the feto-placental unit [183,184]. The improvement in attention scores, adaptability to new surroundings, mental development, memory performance, and hand–eye coordination are associated with higher maternal DHA delivery to the fetal brain [7,185].

DHA deficiency during pregnancy suggests the lower development of language learning skills in children [186]. Even autistic spectrum disorder or ADHD among teenagers is associated with DHA deficiency [187,188]. Neurocognitive functional insufficiency in young adults or loneliness-related memory problems in middle age have been associated with DHA deficiency [189,190]. DHA deficiency in the third trimester significantly causes preterm brain development due to the insufficient maternal consumption of n-3 fatty acids. Even following delivery, infants are entirely dependent on breast milk or formula milk for DHA and ARA. Reduced DHA consumption during this critical brain development period may influence brain functionalities in adult life [191].

Dementia has shown an inverse relationship with regular marine fish consumption in different continents [120,192] and higher blood DHA levels inversely related to dementia [193,194]. Various brain diseases/disorders, such as Alzheimer's disease, Parkinson's disease, Huntington's disease, schizophrenia, and mood disorders, are related to disturbed fatty acid signaling [195,196]. The most prevalent dementia is Alzheimer's disease, which is inversely related to brain DHA level. Serum DHA level reduces significantly, and its addition positively correlates with memory scores in elderly Alzheimer patients [119,197,198]. The plasma DHA level is significantly associated with the risk of Alzheimer's disease [199].

Intake of 200 mg of DHA-containing fish per week reduces the risk of AD by 60% [200]. However, different randomized controlled trials (RCTs) found mixed results with DHA supplementation. DHA alone or combined with ARA or the EPA found no significant neuropsychiatric status changes in Alzheimer's disease patients [123,124,201]. The neurodegenerative disorder, Parkinson's disease's etiology is unknown. However, its primary palliative treatment is dopamine-based therapy. Various animal studies found the neuroprotective effect of DHA in the Parkinson's disease model. DHA has been shown to improve L-DOPA-induced dyskinesia [202] and reduce dopaminergic neuron apoptosis in mouse models [121]. DHA supplements offer a beneficial neuroprotective effect for Parkinson's disease management [203].

Multiple factors influence serotonin biosynthesis and function. The brain's serotonin level correlates with various behavioral consequences, e.g., control of executive function, sensory gating, social behavior, and impulsivity [204]. Serotonin-related gene polymorphism is associated with mental illnesses, e.g., autism spectrum disorders, ADHD, bipolar disorder, schizophrenia, etc. DHA modulates the activity of serotonin in the brain. DHA increases serotonin receptor accessibility by increasing membrane fluidity in postsynaptic neurons [204]. Concentric serotonin and a low DHA level in the orbitofrontal cortex are correlated with schizophrenia [205].

Observational studies showed that ADHD also has a relationship with DHA levels. RCTs of DHA with EPA supplementation and the addition of medications have demonstrated significant improvement in ADHD symptoms [122,187,206–216]. However, DHA supplementation with methylphenidate did not improve ADHD symptoms [217].

There are mixed results in RCTs that have been observed in early psychosis symptoms improvement with DHA supplementation. When DHA is supplemented with EPA for at least 12 weeks, the functional improvement and reduction in psychiatric symptoms are visible in different studies [218–220]. DHA deficiency elicits the chances of schizophrenia by promoter hypermethylation of nuclear receptor genes RxR and PPAR, which results in the downregulation of the gamma-aminobutyric acid-ergic system and the prefrontal cortex involved in oligodendrocyte integrity [221]. Lower erythrocyte DHA status is associated with the development of bipolar disorder [222]. However, combined DHA and EPA supplementation for 6 weeks improved mania and depression among juvenile patients [223–225]. However, recent RCTs found DHA has no role in improving the symptoms of bipolar disorder [20].

7. Can DHA Supplementation Improve Brain Function of Infants: Results of Clinical Trials

Various RCTs proved significant effects of DHA supplementation on infant brain development in pregnancy. The meta-analytic study in 2007 found a static co-relation between visual growth and DHA supplementation in first year of life [125]. An RCT in 2011 showed DHA-supplemented 18-month-old children had higher index scores in mental development [8]. Other RCTs showed DHA-enriched fish oil supplemented children had significantly higher percentile ranks of the total number of gestures at 1 to 1.5 years of age [226]. DHA-formula-fed infants scored equal visual equity scores with breast-fed infants at the age of four [227]. The recent meta-analytic review also found the positive effects of seafood consumption in pregnancies in developing childhood neurocognitive function [53]. The supplementation of DHA to pregnant and nursing mothers and the first year of infant life have developed better cognitive ability. During the last trimester of pregnancy, fetal brain development demands a higher amount of DHA, which can be interrupted in the case of a preterm born baby and can result in mental growth retardation. Different RCTs showed that higher DHA-enriched formula (around 1% of total fatty acids) is essential for the preterm baby for mental growth and development [24].

DHA supplemented along with EPA enhance the outcome for cognitive and mood disorders. However, conflicting data exist about the effect of DHA supplementation on cognition during childhood. The earlier RCT conducted in Australia and Indonesia did not show any improvement in general intelligence or attention among 6–10-year-old children following 88 mg/d DHA supplementation [228]. The DHA Oxford Learning and Behavior (DOLAB) study showed a significant improvement in brain function among aged 7–9-year-olds [229]. Another RCT showed improved memory and learning ability among 7–9-year-old children [230]. A few years ago, another RCT from Australia found no significant difference in academic performance between the DHA-supplemented and control groups [231]. Although the authors have mentioned their study's limitation that the final assessment was finished by less than half of the study population, DHA supplementation was low. The study was conducted with slightly older children, and in some cases, mothers did not give consent or provide data. The Third National Health and Nutrition Examination Survey (NHANES III) found a higher DHA supplementation effect in girls than boys, despite both sexes' receiving cognitive benefit [232]. The pregnant women need higher DHA supplementation for their own growth and for the growth and development of the newborn. Even the source of DHA and the ratio of EPA and DHA may influence the bioavailability of DHA.

A summary of a few recent clinical trials has been shown in the following Table 2. Although a slight improvement, these studies did not show any significant positive outcome from DHA supplementation effects on young children. However, a meta-analysis of DHA supplementation with the EPA improved childhood visual and psychomotor development without significant global IQ effects later in childhood [233]. However, there are no clinical data available where DHA supplementation was conducted before 14.5 weeks of the gestational period.

Table 2. Various clinical studies of DHA supplementation in mothers and infants about brain function.

Study Name	Experimental Setting	Observed Outcome
The Kansas University DHA outcome study (KUDOS) clinical trial	Cognitive and behavioral development	Improvement of visual attention among infants has been observed to reduce the preterm birth risk [234].
Effect of DHA supplementation vs. placebo on developmental outcomes of toddlers born preterm	Developmental outcomes of toddlers	Daily supplementation of DHA did not improve cognitive function and may adversely affect language development and effortful control in specific subgroups of children [235].
Effect of DHA supplementation during pregnancy on maternal depression and neurodevelopment of young children	Neurodevelopmental outcome of children	DHA supplementation during pregnancy did not reduce postpartum depression in mothers, neither did it improve cognitive and language development in their offspring during early childhood [236].
Neurodevelopmental outcomes of preterm infants fed high-amount DHA	Neurodevelopment at 18 months of age	Bayley mental development index scores of preterm infants overall born earlier than 33 weeks were not affected but improved the girls' Bayley mental development index scores.
Neurodevelopmental outcomes at 7 years corrected age in preterm infants who were fed high-dose DHA to term equivalent	Cognitive outcome detected at 18 months age	No evidence of benefit [237].
Feeding preterm infant milk with a higher dose of DHA than that used in current practice	Language or behavior in early childhood	No clinically meaningful change to language development or behavior were observed when assessed in early childhood [238].

8. Roles of Arachidonic acid20:4n-6 (ARA) in Brain Development and Function

In addition to DHA, the mother preferentially supplies ARA to the growing and developing brain via the placenta and breastfeeding. ARA uptake was found to be higher in early trimester trophoblast cells than EPA and DHA [239]. The ARA metabolism in the brain is suspected of having an altered profile in neurological, neurodegenerative, and psychiatric disorders. Using various knock-out models for enzymes involved in brain ARA metabolism, Bosetti showed that the ARA and its metabolites play a significant role in brain physiology via the PLA2/COX pathway [240]. The effects of DHA and ARA on body growth and brain functions were studied using delta-6-desaturase knock-out

(D6D-KO) mice by feeding different combinations of PUFAs in milk formulations. The in vivo findings confirmed the complementary roles of ARA and DHA in body and brain development, respectively [241]. ARA may be required in a higher amount to support growth-promoting placental activities and the production of eicosanoids. In human milk, the amount of ARA typically exceeds the levels of DHA. Milk ARA content is also less varied than DHA, and, unlike DHA, ARA does not seem to be linked to maternal intake. There has been much discussion in recent years about the need for ARA and DHA in infant formula. Studies clearly show the requirement for both ARA and DHA in addition to the essential fatty acids (linoleic acid,18:2n-6 (LA), and ALA to support the optimal body, brain growth, and brain function. ARA is quantitatively the most common LCPUFA in the brain after DHA [242]. Although diverse roles of DHA are investigated, the roles of ARA in brain development and functions have not been investigated to a greater extent. Given that ARA and its precursor, LA, contribute significantly to the Western diet and its pleiotropic biological effects and its interactions with DHA, this n-6 LCPUFA is a crucial modifiable factor in brain development and preventive strategies of brain diseases. ARA corresponds to around 20% of the total amount of neuronal fatty acids and is mainly esterified in membrane phospholipids.

Several studies have suggested that the structure–function and metabolism of the brain depend on levels of ARA and DHA and interactions of their metabolites [29]. The recycling (de-esterification–re-esterification) of these two fatty acids in the brain are independently carried out by ARA- and DHA-selective enzymes. The ARA-mediated processes can be targeted or altered separately from the DHA-mediated processes by a dietary deficiency of n-3 PUFA or genetic manipulation. Therefore, in studies using n-3 PUFA deficiency models, the homeostatic mechanisms show DHA loss in the brain while increasing ARA metabolism. Further studies are required to understand the impact of the n-6/n-3 ratio on the regulation of DHA-selective iPLA2 and COX-1 or ARA-selective cPLA2, sPLA2, and COX-2 and their effects on brain function and neuroinflammation.

ARA must either be consumed in the diet or synthesized from its precursor LA in the liver. The brain contains relatively low LA levels, and its conversion into ARA is minimum in the brain. Thus, the growing brain depends on a steady supply of ARA [243] from the maternal circulation or via breast milk. Although lipoproteins and lysophospholipids of plasma may contribute to brain ARA levels, their quantitative contribution is unknown. Upon its entry into the brain, ARA is activated by a long-chain acyl-CoA synthetase and can be esterified into the sn-2 position of phospholipids. During neurotransmission, the brain ARA cascade is initiated when ARA is released from synaptic membrane phospholipid by the neuroreceptor-initiated activation of cPLA$_2$. PLA$_2$ is activated by dopaminergic, cholinergic, glutamatergic, and serotonergic stimulation via G-proteins or calcium [244]. Several PLA$_2$ are activated via serotonergic (5-hydroxytryptaminergic), glutamatergic, dopaminergic, and cholinergic receptors [244,245]. Usually, calcium-dependent cytosolic PLA$_2$ (cPLA$_2$) resides at the postsynaptic terminals, selective for releasing ARA, whereas calcium-independent PLA$_2$ is believed to release the DHA sn-2 position of phospholipids [13,14]. Upon its release, a portion of the unesterified ARA is converted to prostaglandins, leukotrienes, and lipoxins, a portion oxidized via β-oxidation, and the remainder (approximately 97% under basal conditions) is activated by ACSL and ultimately recycled and re-esterified into the sn-2 position of phospholipids [246]. An additional ARA is released by activated cytokine and glutamatergic N-methyl-d-aspartate receptors in conditions such as neuroinflammation and excitotoxicity. Although the signals that ARA and its derivatives relay are not entirely understood, they regulate blood flow, neuroinflammation, excitotoxicity, the sleep/wake cycle, and neurogenesis [247].

Like DHA, ARA is also directly involved in synaptic functions. The level of intracellular free ARA and the balance between the releasing and incorporating enzymes in membrane phospholipids may play critical roles in neuroinflammation and synaptic dysfunction. Both these events are observed in the murine model of Alzheimer's disease before the amyloid plaques and the neurofibrillary tangles, respectively formed by the two agents known for Alzheimer's disease agents, A β peptide and hyperphosphorylated *tau*. Finally, western food, which contains excessive n-6/n-3 ratios, might favor more ARA levels and influence Alzheimer's disease mechanisms.

A better understanding of the complex relationships between ARA and DHA and their brain mechanisms is required. Free ARA contributes to Alzheimer's disease progression via different pathways. ARA and derivatives are pro-inflammatory and participate in neuroinflammation. ARA is directly involved in synaptic functions as a retrograde messenger and a regulator of neuro mediator exocytosis. ARA also influences tau phosphorylation, and polymerization can compete with DHA. Moreover, ARA has pleiotropic effects on brain disease, and it may be used in the fight against brain diseases. The dietary ARA and brain diseases about DHA should be investigated further to prevent the disease.

The transgenic Alzheimer's disease murine model showed that dietary ARA produced opposite $A\beta$ production in Alzheimer's disease. Studies on the impact of dietary ARA on Alzheimer's disease are required to identify the underlying mechanisms of action. A reduced level of ARA in the temporal cortex of Alzheimer's patients was observed [248]; however, its relation with DHA and its metabolites is unknown. ARA involvement in Alzheimer's disease was mediated via $cPLA_2\alpha$. However, a DHA-rich diet did not show such effects [249]. A diet containing 2% ARA for 21 weeks increased $A\beta_{1-42}$ production and deposition in 24-week-old CRND8 mice [250].

ARA is involved in cell division and signaling during brain growth and development [251]. In addition, ARA mediates neuronal firing [252], signaling [253] and long-term potentiation [254]. The absolute levels of n-3 PUFAs and the ratio of n-6 and n-3 PUFA affect gene expression of controlling neurogenesis and neural function.

ARA maintains structural order of the membrane and hippocampal plasticity [255]. ARA also protects the brain against oxidative stress in the hippocampus region via PPARγ and synthesizes new proteins [256]. Released intracellular ARA activates protein kinases and ion channels, inhibits the uptake of neurotransmitters, and enhances synaptic transmission, and modulates neuronal excitability [251]. As ARA is involved in intracellular signaling, the optimum levels of intracellular free ARA must be steadily maintained. ARA also activates syntaxin-3 (STX-3), a plasma membrane protein involved in the growth and repair of neurites [257]. Neurite growth closely correlates with the ARA-mediated activation of STX-3 in membrane expansion at growth cones [257]. The neurite growth from the cell body is a critical step in neuronal development. ARA stimulates exocytosis by allowing the attachment of STX-3 with the fusogenic soluble N-ethylmaleimide-sensitive factor receptors (SNARE complex) [258]. The SNARE proteins are involved in producing a fusion of vesicular and plasma membranes in the brain. The formation of this SNARE complex mediated by ARA drives membrane fusion, leading to the release of vesicular cargo into the extracellular spaces [258]. α-Synuclein plays a role in the development of Parkinson's disease, can sequester ARA and thus blocks the activation of the SNARE complex [258], suggesting the importance of ARA in synaptic transmission. All these data show the importance of ARA in cell signaling, trafficking, and the regulation of spatial–temporal interactions between cellular structures.

The most abundant prostaglandins in the brain are PGD_2 and PGE_2. They are synthesized from ARA by PGD_2 and PGE_2 synthases. COX-2 is overexpressed in the cortex and hippocampus of Alzheimer's disease patients [259]. PGE_2 increases neuroinflammation and amplifies Alzheimer's disease pathology through various mechanisms.

Figure 4 describes different metabolites of ARA involved in brain function. ARA-derived lipoxins are anti-inflammatory eicosanoids distinct from pro-inflammatory leukotriene and prostaglandin. Lipoxin biosynthesis occurs via two different pathways. Lipoxins mediate their action on endothelial cells to offer an inflammation resolution process. LipoxinA4 lowers neuroinflammation by reducing microglial activation. 5- Lipoxygenase (5-LOX) converts ARA into 5-HPETE and then 5-HETE or leukotriene A4 (LTA4). There is the increased expression of the dual enzyme 12/15-LOX and its products.

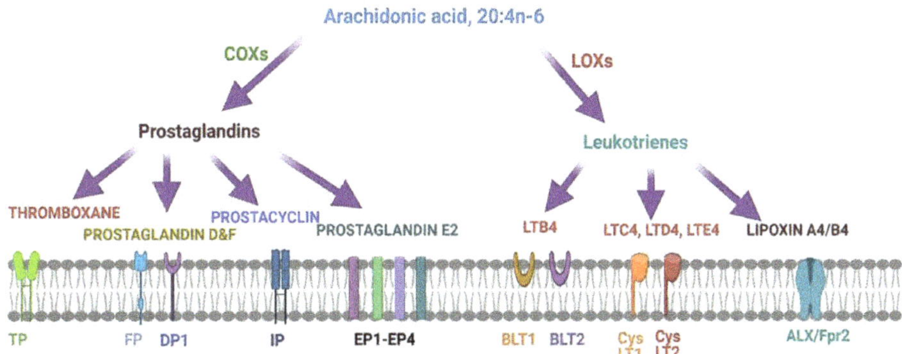

Figure 4. Metabolites of ARA and their receptors.

ARA is converted to lipoxinA4 or lipoxinB4 via two different pathways. Lipoxins mediate their action on endothelial cells to offer an inflammation resolution process. LipoxinA4 lowers neuroinflammation by reducing microglial activation. LTA4, LTB4, LTC4, LTD4, and LTE4 are leukotrienes A4, B4, C4, D4, and E4, respectively; DP1, EP1-EP4, FP, and IP, prostaglandin receptors; TP, thromboxane A2 receptor; BLT1 and BLT2, leukotriene B4 receptor; CysLT1 and CysLT2: cysteinyl leukotriene receptors; ALXR: lipoxins receptor.

12(S)-HETE and 15(S)-HETE were reported in frontal and temporal brain regions' cerebrospinal fluid, respectively, in Alzheimer's disease patients.

The endocannabinoids containing ARA are involved in Alzheimer's disease via the CB2 receptor. Although neuroinflammation is related to Alzheimer's disease, there is no evidence that higher brain contents of ARA would produce inflammation.

Several animal studies indicate that ARA has beneficial effects on cognition and synaptic plasticity [260]. An association between abilities of spatial memory and ARA content of the hippocampus was reported. ARA modulates Kv channels at the postsynaptic membrane, and influences long-term potentiation [261]. ARA induces presynaptic long-term depression associated with a Ca^{2+} influx and the activation of metabotropic glutamate receptors [261]. In addition, ARA can induce neurotransmitter exocytosis in the presynaptic neuron via activation of the soluble N-ethylmaleimide-attachment receptors (SNARE). ARA induces the binding of syntaxin-1 to the SNARE complex in the presence of Munc18-1, which is a critical regulator of the process [262]. Although ARA's role in synaptic function is well documented, the benefit or drawback may depend on the delicate balance of ARA levels. Age and Aβ concentration or other physiological factors could modify ARA's effects on synapse and neuronal cells. Additional work is now required to characterize dietary ARA's influence on associated brain diseases.

9. Transport of ARA to the Developing Brain

The transport of ARA to the brain tissue and inside the brain are not well understood, and more investigations are required for their nuanced characterization. The biological roles of several ACSLs and LPATs for incorporating ARA in the membrane phospholipids have been studied [246]. Some of them are relatively specific for ARA, such as ACLS4, LPIAT1, and LPCAT3. The polymorphisms and variations in their expression levels can affect ARA deposition in the different tissues, including the brain, despite identical ARA content in the food. Nutritional recommendations regarding the ARA intake must be based on the knowledge of such variations.

ARA's effects on brain signaling functions depend on the fact that free ARA is released from membrane phospholipid by cPLA$_2\alpha$ or secretory PLA$_2$. The turnover of ARA in brain membrane phospholipids is involved bipolar disorder. The downregulation of the turnover correlates with the reduced expression of ARA-selective cPLA2 or acyl-CoA synthetase and COX-2 [244].

It is not known whether higher ARA amounts in specific phospholipid classes in the brain favor ARA released by these PLA_2 and lead to higher free ARA levels in neuronal or glial cells. Further studies are required to determine the effect of ARA brain uptake on its release and conversion into pro- or anti-inflammatory derivatives.

While the competition between n-6 and n-3 fatty acids and their precursors for the desaturases, COXs, and LOXs are known, little is known on the competition between ARA and DHA in incorporation in the brain. Mfsd2A transports DHA to the brain in the form of DHA-LPC [74], but brain ARA content was increased by 30% in Mfsd2A knockout mice, whereas that of DHA was reduced by 58%. These fatty acid changes were accompanied by neuronal loss in the hippocampus and cerebellum with severe cognitive deficits and anxiety. Therefore, the deposition of ARA to the brain is not altered by the reduced expression of Mfsd2A. However, under this circumstance, ARA replaced DHA in phospholipids but could not play the roles of DHA in neuronal functions [74].

FABPs also have different affinities for ARA and DHA. FABP5 and FABP7 are more selective for DHA, whereas FABP3 binds ARA with much higher affinity [263,264]. Regarding fatty acid metabolizing enzymes, DHA is preferentially used as a substrate by ACSL6 [265] and calcium-independent group VI PLA_2 [266], whereas ACSL4 and group IV $cPLA_2$ used ARA as the preferred substrate.

ARA and DHA are preferentially incorporated in different lipid fractions of the brain regions, suggesting a differential role of these lipid biomolecules. ARA is concentrated into brain inositol phospholipid (PL), ethanolamine plasmalogens lipid fractions, while DHA is accumulated in serine and ethanolamine PL fractions of the synaptic membranes. ARA-rich PL is more enriched in white matter (myelinated regions) than grey matter, and there are regional differences in ARA and DHA in the brain. The DHA, ARA, and adrenic acid (22:4n-6) are the most abundant PUFAs in the brain. The preterm infants had less DHA and a lower DHA/ARA ratio in both the brain and the retina than term infants. [267] The mean proportions of ARA were higher in the early placenta than term, while its immediate precursor dihomo-gamma-linolenic acid,20:3n-6 (DGLA) was higher in term than the early placenta. The increased presence of ARA during early placentation supports its organogenesis and vascularization activities. In contrast, the enhanced proportions of DGLA favors the optimal blood flow to sustain fetal growth by their vasorelaxant and anti-platelet effects of PGE_1-like activities [268,269]. ARA and DHA are required to replenish brain injury, vascular regulation, and brain development in preterm babies. ARA is needed to support endothelial cells during brain injury, while DHA is required to support membrane fluidity for the receptors and the linear growth and network of neuronal cells. The neuroprotective actions of DHA in preterm are manifested with anti-inflammatory initiation and resolutions by its own and its signaling lipid mediators in particular. ARA and DHA produced different sets of pro-resolving lipid bioactive. Lipoxins are derived from ARA, while resolvin (neuro) protectin and maresins are derived from EPA and DHA. LXA4 receptor staining in the brain indicates that lipoxinA4 lowers neuroinflammation and brain edema during brain injury [270]. The anti-inflammatory lipoxinA4 acts as an endogenous allosteric modulator of the cannabinoid receptor [271]. Thus, in contrast to PGE2, which shows pro-inflammatory actions, lipoxins offer an inflammation resolution process.

10. Conclusions

The essentiality of both DHA and ARA is known by the fact that these fatty acids make up 20% of the brain's fatty acids. The mother is required to supply these two fatty acids preferentially during the critical period of brain development via the placenta in utero and breastfeeding postnatally. However, the data on maternal and infant nutritional intakes are not yet consistent, despite the fact that DHA's impact on brain development and function has been investigated extensively for the last two decades. Due to some compounded variables and conflicting reports on the association with LCPUFA supplementation and cognitive development, it imperative to determine optimum DHA doses in the presence of ARA for optimum brain development in infants [272–274].

The mean concentrations of DHA and ARA in breast milk can vary based on the maternal diet [260]. The group with the highest levels of DHA showed decreased ARA levels in two brain areas, suggesting the competition of DHA with ARA. Still, the absolute levels of ARA and DHA could be more important than their ratio, in particular with the preterm infant. ARA supports the first year of developmental growth as the conditional deficiency of ARA is established in preterm infants [60].

The optimal balance of DHA and ARA intake during infancy is still unknown, but the current best practice advises that the amount of DHA in infant formula should not exceed the amount of ARA [275].

The high doses of DHA in formula may suppress the benefits of ARA. The preterm infant who received formula with the n-6/n-3 ratio of 2:1 showed a higher level of LCP-UFAs and improved psychomotor development than the n-6/n-3 ratio of the 1:1 group, suggesting an ARA and DHA ratio of 2:1 in the formula for the very preterm infant [276]. Thus, the common requirement for nutritional supports needs both ARA and DHA for vascular and neuronal development, in particular with pregnancies where the supply of these nutrients halted prematurely [277]. The supplementation of DHA:ARA in infant formulas ranges from 1:1 to 1:2. The present recommendations with DHA and ARA levels are that 0.2% to 0.4% and 0.35% to 0.7% of total fatty acids are appropriate [278]. The formula should reflect human milk composition for optimal neurocognitive benefits. Both ARA and DHA are necessary for fetal neurodevelopment, and a deficiency in one may compromise growth. However, further works are required to understand the complementary roles of ARA and DHA in neurodevelopment.

Author Contributions: Conceptualization, A.K.D.; Methodology, A.K.D. and S.B.; Investigation, A.K.D. Original Draft Preparation A.K.D.; Writing—Review and Editing, A.K.D., S.B., R.M.; A.B., S.P.; Visualization, R.M., S.B. and A.K.D.; Supervision, A.K.D. All authors have read and agreed to the published version of the manuscript.

Funding: This work is supported in part by the grants from Throne Holst Foundation and the Faculty of Medicine, University Oslo, Norway.

Conflicts of Interest: The authors express no conflict of interest.

Abbreviations

ARA: arachidonic acid; AD: Alzheimer's disease; ADHD: attention deficit hyperactivity disorder; ALA: α-linolenic acid, 18:3n-3; COX: cyclooxygenase; DHA: docosahexaenoic acid,22:6n-3; EPA: eicosapentaenoic acid,20:5n-3; LCPUFAs: long-chain polyunsaturated fatty acids; LOX: lipoxygenases; NO: nitric oxide; PUFAs: polyunsaturated fatty acids; RCT: randomized controlled trial.

References

1. Nyaradi, A.; Li, J.; Hickling, S.; Foster, J.; Oddy, W.H. The role of nutrition in children's neurocognitive development, from pregnancy through childhood. *Front. Hum. Neurosci.* **2013**, *7*, 97. [CrossRef]
2. Lauritzen, L.; Brambilla, P.; Mazzocchi, A.; Harslof, L.B.; Ciappolino, V.; Agostoni, C. DHA Effects in brain development and function. *Nutrients* **2016**, *8*, 6. [CrossRef]
3. Martinez, M.; Conde, C.; Ballabriga, A. Some chemical aspects of human brain development. II. Phosphoglyceride fatty acids. *Pediatr. Res.* **1974**, *8*, 93–102. [CrossRef]
4. Clandinin, M.T.; Chappell, J.E.; Leong, S.; Heim, T.; Swyer, P.R.; Chance, G.W. Intrauterine fatty acid accretion rates in human brain: Implications for fatty acid requirements. *Early Hum. Dev.* **1980**, *4*, 121–129. [CrossRef]
5. Agostoni, C.; Trojan, S.; Bellu, R.; Riva, E.; Giovannini, M. Neurodevelopmental quotient of healthy term infants at 4 months and feeding practice: The role of long-chain polyunsaturated fatty acids. *Pediatr. Res.* **1995**, *38*, 262–266. [CrossRef] [PubMed]
6. Agostoni, C.; Trojan, S.; Bellu, R.; Riva, E.; Bruzzese, M.G.; Giovannini, M. Developmental quotient at 24 months and fatty acid composition of diet in early infancy: A follow up study. *Arch. Dis. Child* **1997**, *76*, 421–424. [CrossRef] [PubMed]
7. Helland, I.B.; Smith, L.; Saarem, K.; Saugstad, O.D.; Drevon, C.A. Maternal supplementation with very-long-chain n-3 fatty acids during pregnancy and lactation augments children's IQ at 4 years of age. *Pediatrics* **2003**, *111*, e39–e44. [CrossRef] [PubMed]

8. Drover, J.R.; Hoffman, D.R.; Castañeda, Y.S.; Morale, S.E.; Garfield, S.; Wheaton, D.H.; Birch, E.E. Cognitive function in 18-month-old term infants of the DIAMOND study: A randomized, controlled clinical trial with multiple dietary levels of docosahexaenoic acid. *Early Hum. Dev.* **2011**, *87*, 223–230. [CrossRef]

9. Innis, S.M. Perinatal biochemistry and physiology of long-chain polyunsaturated fatty acids. *J. Pediatr.* **2003**, *143*, S1–S8. [CrossRef]

10. Jensen, C.L.; Voigt, R.G.; Prager, T.C.; Zou, Y.L.; Fraley, J.K.; Rozelle, J.C.; Turcich, M.R.; Llorente, A.M.; Anderson, R.E.; Heird, W.C. Effects of maternal docosahexaenoic acid intake on visual function and neurodevelopment in breastfed term infants. *Am. J. Clin. Nutr.* **2005**, *82*, 125–132. [CrossRef]

11. Dutta-Roy, A.K. Transport mechanisms for long-chain polyunsaturated fatty acids in the human placenta. *Am. J. Clin. Nutr.* **2000**, *71*, 315S–322S. [CrossRef]

12. Basak, S.; Mallick, R.; Duttaroy, A.K. Maternal docosahexaenoic acid status during pregnancy and its impact on infant neurodevelopment. *Nutrients* **2020**, *12*, 3615. [CrossRef]

13. Lauritzen, L.; Carlson, S.E. Maternal fatty acid status during pregnancy and lactation and relation to newborn and infant status. *Matern. Child Nutr.* **2011**, *7*, 41–58. [CrossRef] [PubMed]

14. Browning, L.M.; Walker, C.G.; Mander, A.P.; West, A.L.; Madden, J.; Gambell, J.M.; Young, S.; Wang, L.; Jebb, S.A.; Calder, P.C. Incorporation of eicosapentaenoic and docosahexaenoic acids into lipid pools when given as supplements providing doses equivalent to typical intakes of oily fish. *Am. J. Clin. Nutr.* **2012**, *96*, 748–758. [CrossRef]

15. Innis, S.M.; Friesen, R.W. Essential n-3 fatty acids in pregnant women and early visual acuity maturation in term infants. *Am. J. Clin. Nutr.* **2008**, *87*, 548–557. [CrossRef] [PubMed]

16. Yavin, E.; Himovichi, E.; Eilam, R. Delayed cell migration in the developing rat brain following maternal omega 3 alpha linolenic acid dietary deficiency. *Neuroscience* **2009**, *162*, 1011–1022. [CrossRef] [PubMed]

17. Su, K.P. Biological mechanism of antidepressant effect of omega-3 fatty acids: How does fish oil act as a 'mind-body interface'? *Neurosignals* **2009**, *17*, 144–152. [CrossRef] [PubMed]

18. Su, K.P.; Matsuoka, Y.; Pae, C.U. Omega-3 polyunsaturated fatty acids in prevention of mood and anxiety disorders. *Clin. Psychopharmacol. Neurosci.* **2015**, *13*, 129–137. [CrossRef] [PubMed]

19. Brunst, K.J.; Enlow, M.B.; Kannan, S.; Carroll, K.N.; Coull, B.A.; Wright, R.J. Effects of prenatal social stress and maternal dietary fatty acid ratio on infant temperament: Does race matter? *Epidemiology* **2014**, *4*. [CrossRef]

20. Ciappolino, V.; DelVecchio, G.; Prunas, C.; Andreella, A.; Finos, L.; Caletti, E.; Siri, F.; Mazzocchi, A.; Botturi, A.; Turolo, S.; et al. The Effect of DHA supplementation on cognition in patients with bipolar disorder: An exploratory randomized control trial. *Nutrients* **2020**, *12*, 708. [CrossRef]

21. Clayton, E.H.; Hanstock, T.L.; Hirneth, S.J.; Kable, C.J.; Garg, M.L.; Hazell, P.L. Long-chain omega-3 polyunsaturated fatty acids in the blood of children and adolescents with juvenile bipolar disorder. *Lipids* **2008**, *43*, 1031–1038. [CrossRef]

22. Hahn-Holbrook, J.; Fish, A.; Glynn, L.M. Human milk omega-3 fatty acid composition is associated with infant temperament. *Nutrients* **2019**, *11*, 2964. [CrossRef] [PubMed]

23. Grey, K.R.; Davis, E.P.; Sandman, C.A.; Glynn, L.M. Human milk cortisol is associated with infant temperament. *Psychoneuroendocrinology* **2013**, *38*, 1178–1185. [CrossRef]

24. Weiser, M.J.; Butt, C.M.; Mohajeri, M.H. Docosahexaenoic acid and cognition throughout the lifespan. *Nutrients* **2016**, *8*, 99. [CrossRef]

25. Lin, P.Y.; Su, K.P. A meta-analytic review of double-blind, placebo-controlled trials of antidepressant efficacy of omega-3 fatty acids. *J. Clin. Psychiatry* **2007**, *68*, 1056–1061. [CrossRef] [PubMed]

26. Bhatia, H.S.; Agrawal, R.; Sharma, S.; Huo, Y.X.; Ying, Z.; Gomez-Pinilla, F. Omega-3 fatty acid deficiency during brain maturation reduces neuronal and behavioral plasticity in adulthood. *PLoS ONE* **2011**, *6*, e28451. [CrossRef]

27. Mallick, R.; Basak, S.; Duttaroy, A.K. Docosahexaenoic acid,22:6n-3: Its roles in the structure and function of the brain. *Int. J. Develop. Neurosci. Off. J. Int. Soc. Develop. Neurosci.* **2019**, *79*, 21–31. [CrossRef] [PubMed]

28. Duvall, M.G.; Levy, B.D. DHA- and EPA-derived resolvins, protectins, and maresins in airway inflammation. *Eur. J. Pharmacol.* **2016**, *785*, 144–155. [CrossRef] [PubMed]

29. Contreras, M.A.; Rapoport, S.I. Recent studies on interactions between n-3 and n-6 polyunsaturated fatty acids in brain and other tissues. *Curr. Opin. Lipidol.* **2002**, *13*, 267–272. [CrossRef]

30. Colombo, J.; Carlson, S.E.; Cheatham, C.L.; Shaddy, D.J.; Kerling, E.H.; Thodosoff, J.M.; Gustafson, K.M.; Brez, C. Long-term effects of LCPUFA supplementation on childhood cognitive outcomes. *Am. J. Clin. Nutr.* **2013**, *98*, 403–412. [CrossRef]

31. Lepping, R.J.; Honea, R.A.; Martin, L.E.; Liao, K.; Choi, I.Y.; Lee, P.; Papa, V.B.; Brooks, W.M.; Shaddy, D.J.; Carlson, S.E.; et al. Long-chain polyunsaturated fatty acid supplementation in the first year of life affects brain function, structure, and metabolism at age nine years. *Dev. Psychobiol.* **2019**, *61*, 5–16. [CrossRef] [PubMed]

32. McNamara, R.K.; Carlson, S.E. Role of omega-3 fatty acids in brain development and function: Potential implications for the pathogenesis and prevention of psychopathology. *Prostaglandins Leukot. Essent. Fatty Acids* **2006**, *75*, 329–349. [CrossRef] [PubMed]

33. Montes, R.; Chisaguano, A.M.; Castellote, A.I.; Morales, E.; Sunyer, J.; Lopez-Sabater, M.C. Fatty-acid composition of maternal and umbilical cord plasma and early childhood atopic eczema in a Spanish cohort. *Eur. J. Clin. Nutr.* **2013**, *67*, 658–663. [CrossRef] [PubMed]

34. Kuipers, R.S.; Luxwolda, M.F.; Offringa, P.J.; Boersma, E.R.; Dijck-Brouwer, D.A.; Muskiet, F.A. Fetal intrauterine whole body linoleic, arachidonic and docosahexaenoic acid contents and accretion rates. *Prostaglandins Leukot. Essent. Fatty Acids* **2012**, *86*, 13–20. [CrossRef]
35. Innis, S.M. Essential fatty acids in growth and development. *Prog. Lipid Res.* **1991**, *30*, 39–103. [CrossRef]
36. Martinez, M.; Mougan, I. Fatty acid composition of human brain phospholipids during normal development. *J. Neurochem.* **1998**, *71*, 2528–2533. [CrossRef] [PubMed]
37. Salem, N., Jr.; Wegher, B.; Mena, P.; Uauy, R. Arachidonic and docosahexaenoic acids are biosynthesized from their 18-carbon precursors in human infants. *Proc. Natl. Acad. Sci. USA* **1996**, *93*, 49–54. [CrossRef]
38. Dijck-Brouwer, D.A.; Hadders-Algra, M.; Bouwstra, H.; Decsi, T.; Boehm, G.; Martini, I.A.; Boersma, E.R.; Muskiet, F.A. Lower fetal status of docosahexaenoic acid, arachidonic acid and essential fatty acids is associated with less favorable neonatal neurological condition. *Prostaglandins Leukot. Essent. Fatty Acids* **2005**, *72*, 21–28. [CrossRef]
39. Sherry, C.L.; Oliver, J.S.; Marriage, B.J. Docosahexaenoic acid supplementation in lactating women increases breast milk and plasma docosahexaenoic acid concentrations and alters infant omega 6:3 fatty acid ratio. *Prostaglandins Leukot. Essent. Fatty Acids* **2015**, *95*, 63–69. [CrossRef]
40. Cetin, I.; Alvino, G.; Cardellicchio, M. Long chain fatty acids and dietary fats in fetal nutrition. *J. Physiol.* **2009**. [CrossRef]
41. Duttaroy, A.K. Transport of fatty acids across the human placenta: A review. *Prog. Lipid Res.* **2009**, *48*, 52–61. [CrossRef]
42. Innis, S.M. Impact of maternal diet on human milk composition and neurological development of infants. *Am. J. Clin. Nutr.* **2014**, *99*, 734S–741S. [CrossRef] [PubMed]
43. Hsieh, A.T.; Brenna, J.T. Dietary docosahexaenoic acid but not arachidonic acid influences central nervous system fatty acid status in baboon neonates. *Prostaglandins Leukot. Essent. Fatty Acids* **2009**, *81*, 105–110. [CrossRef]
44. Calder, P.C. The DHA content of a cell membrane can have a significant influence on cellular behaviour and responsiveness to signals. *Ann. Nutrit. Metab.* **2016**, *69*, 8–21. [CrossRef]
45. Basak, S.; Das, M.K.; Duttaroy, A.K. Fatty acid-induced angiogenesis in first trimester placental trophoblast cells: Possible roles of cellular fatty acid-binding proteins. *Life Sci.* **2013**, *93*, 755–762. [CrossRef] [PubMed]
46. Johnsen, G.M.; Basak, S.; Weedon-Fekjaer, M.S.; Staff, A.C.; Duttaroy, A.K. Docosahexaenoic acid stimulates tube formation in first trimester trophoblast cells, HTR8/SVneo. *Placenta* **2011**, *32*, 626–632. [CrossRef]
47. Khong, Y.; Brosens, I. Defective deep placentation. *Best Pract. Res. Clin. Obstet. Gynaecol.* **2011**, *25*, 301–311. [CrossRef]
48. Campbell, F.M.; Dutta-Roy, A.K. Plasma membrane fatty acid-binding protein (FABPpm) is exclusively located in the maternal facing membranes of the human placenta. *FEBS Lett.* **1995**, *375*, 227–230. [CrossRef]
49. Bourre, J.M. Dietary omega-3 fatty acids for women. *Biomed. Pharmacother.* **2007**, *61*, 105–112. [CrossRef]
50. Jorgensen, M.H.; Nielsen, P.K.; Michaelsen, K.F.; Lund, P.; Lauritzen, L. The composition of polyunsaturated fatty acids in erythrocytes of lactating mothers and their infants. *Matern. Child. Nutr.* **2006**, *2*, 29–39. [CrossRef] [PubMed]
51. Koletzko, B.; Schmidt, E.; Bremer, H.J.; Haug, M.; Harzer, G. Effects of dietary long-chain polyunsaturated fatty acids on the essential fatty acid status of premature infants. *Eur. J. Pediatr.* **1989**, *148*, 669–675. [CrossRef]
52. Innis, S.M. Chapter 10 Essential fatty acid metabolism during early development. *Biol. Grow. Anim.* **2005**. [CrossRef]
53. Hibbeln, J.R.; Spiller, P.; Brenna, J.T.; Golding, J.; Holub, B.J.; Harris, W.S.; Kris-Etherton, P.; Lands, B.; Connor, S.L.; Myers, G.; et al. Relationships between seafood consumption during pregnancy and childhood and neurocognitive development: Two systematic reviews. *Prostaglandins Leukot. Essent. Fatty Acids* **2019**, *151*, 14–36. [CrossRef]
54. Oddy, W.H.; Kendall, G.E.; Li, J.; Jacoby, P.; Robinson, M.; de Klerk, N.H.; Silburn, S.R.; Zubrick, S.R.; Landau, L.I.; Stanley, F.J. The long-term effects of breastfeeding on child and adolescent mental health: A pregnancy cohort study followed for 14 years. *J. Pediatr.* **2010**, *156*, 568–574. [CrossRef] [PubMed]
55. Marangoni, F.; Agostoni, C.; Lammardo, A.M.; Giovannini, M.; Galli, C.; Riva, E. Polyunsaturated fatty acid concentrations in human hindmilk are stable throughout 12-months of lactation and provide a sustained intake to the infant during exclusive breastfeeding: An Italian study. *Br. J. Nutr.* **2000**, *84*, 103–109. [CrossRef]
56. Brenna, J.T.; Varamini, B.; Jensen, R.G.; Diersen-Schade, D.A.; Boettcher, J.A.; Arterburn, L.M. Docosahexaenoic and arachidonic acid concentrations in human breast milk worldwide. *Am. J. Clin. Nutr.* **2007**, *85*, 1457–1464. [CrossRef]
57. Makrides, M.; Neumann, M.A.; Gibson, R.A. Effect of maternal docosahexaenoic acid (DHA) supplementation on breast milk composition. *Eur. J. Clin. Nutr.* **1996**, *50*, 352–357. [PubMed]
58. Lattka, E.; Rzehak, P.; Szabo, E.; Jakobik, V.; Weck, M.; Weyermann, M.; Grallert, H.; Rothenbacher, D.; Heinrich, J.; Brenner, H.; et al. Genetic variants in the FADS gene cluster are associated with arachidonic acid concentrations of human breast milk at 1.5 and 6 mo postpartum and influence the course of milk dodecanoic, tetracosenoic, and trans-9-octadecenoic acid concentrations over the duration of lactation. *Am. J. Clin. Nutr.* **2011**, *93*, 382–391. [CrossRef]
59. Van Goor, S.A.; Dijck-Brouwer, D.A.; Hadders-Algra, M.; Doornbos, B.; Erwich, J.J.; Schaafsma, A.; Muskiet, F.A. Human milk arachidonic acid and docosahexaenoic acid contents increase following supplementation during pregnancy and lactation. *Prostaglandins Leukot. Essent. Fatty Acids* **2009**, *80*, 65–69. [CrossRef] [PubMed]
60. Carlson, S.E.; Werkman, S.H.; Peeples, J.M.; Cooke, R.J.; Tolley, E.A. Arachidonic acid status correlates with first year growth in preterm infants. *Proc. Natl. Acad. Sci. USA* **1993**, *90*, 1073–1077. [CrossRef]
61. Uauy-Dagach, R.; Mena, P. Nutritional role of omega-3 fatty acids during the perinatal period. *Clin. Perinatol.* **1995**, *22*, 157–175. [CrossRef]

62. Jimenez, M.J.; Bocos, C.; Panadero, M.; Herrera, E. Fish oil diet in pregnancy and lactation reduces pup weight and modifies newborn hepatic metabolic adaptations in rats. *Eur. J. Nutr.* **2017**, *56*, 409–420. [CrossRef]
63. Simopoulos, A.P. An increase in the omega-6/omega-3 fatty acid ratio increases the risk for obesity. *Nutrients* **2016**, *8*, 128. [CrossRef] [PubMed]
64. Cohen, J.T.; Bellinger, D.C.; Connor, W.E.; Shaywitz, B.A. A quantitative analysis of prenatal intake of n-3 polyunsaturated fatty acids and cognitive development. *Am. J. Prev. Med.* **2005**, *29*, 366–374. [CrossRef] [PubMed]
65. Dunstan, J.A.; Simmer, K.; Dixon, G.; Prescott, S.L. Cognitive assessment of children at age 2(1/2) years after maternal fish oil supplementation in pregnancy: A randomised controlled trial. *Arch. Dis. Child Fetal. Neonatal. Ed.* **2008**, *93*, F45–F50. [CrossRef] [PubMed]
66. Oddy, W.H.; Li, J.; Whitehouse, A.J.; Zubrick, S.R.; Malacova, E. Breastfeeding duration and academic achievement at 10 years. *Pediatrics* **2011**, *127*, e137–e145. [CrossRef]
67. Kramer, M.S.; Aboud, F.; Mironova, E.; Vanilovich, I.; Platt, R.W.; Matush, L.; Igumnov, S.; Fombonne, E.; Bogdanovich, N.; Ducruet, T.; et al. Breastfeeding and child cognitive development: New evidence from a large randomized trial. *Arch. Gen. Psychiatry* **2008**, *65*, 578–584. [CrossRef]
68. Caspi, A.; Williams, B.; Kim-Cohen, J.; Craig, I.W.; Milne, B.J.; Poulton, R.; Schalkwyk, L.C.; Taylor, A.; Werts, H.; Moffitt, T.E. Moderation of breastfeeding effects on the IQ by genetic variation in fatty acid metabolism. *Proc. Natl. Acad. Sci. USA* **2007**, *104*, 18860–18865. [CrossRef] [PubMed]
69. Van Elswyk, M.E.; Kuratko, C. Achieving adequate DHA in maternal and infant diets. *J. Am. Diet Assoc.* **2009**, *109*, 403–404. [CrossRef] [PubMed]
70. Innis, S.M. Dietary omega 3 fatty acids and the developing brain. *Brain Res.* **2008**, *1237*, 35–43. [CrossRef] [PubMed]
71. Kadry, H.; Noorani, B.; Cucullo, L. A blood-brain barrier overview on structure, function, impairment, and biomarkers of integrity. *Fluids Barriers CNS* **2020**, *17*, 69. [CrossRef]
72. Dutta-Roy, A.K. Cellular uptake of long-chain fatty acids: Role of membrane-associated fatty-acid-binding/transport proteins. *Cell. Mol. Life Sci.* **2000**, *57*, 1360–1372. [CrossRef]
73. Mitchell, R.W.; On, N.H.; Del Bigio, M.R.; Miller, D.W.; Hatch, G.M. Fatty acid transport protein expression in human brain and potential role in fatty acid transport across human brain microvessel endothelial cells. *J. Neurochem.* **2011**, *117*, 735–746. [CrossRef] [PubMed]
74. Nguyen, L.N.; Ma, D.; Shui, G.; Wong, P.; Cazenave-Gassiot, A.; Zhang, X.; Wenk, M.R.; Goh, E.L.; Silver, D.L. Mfsd2a is a transporter for the essential omega-3 fatty acid docosahexaenoic acid. *Nature* **2014**, *509*, 503–506. [CrossRef] [PubMed]
75. Guemez-Gamboa, A.; Nguyen, L.N.; Yang, H.; Zaki, M.S.; Kara, M.; Ben-Omran, T.; Akizu, N.; Rosti, R.O.; Rosti, B.; Scott, E.; et al. Inactivating mutations in MFSD2A, required for omega-3 fatty acid transport in brain, cause a lethal microcephaly syndrome. *Nat. Genet.* **2015**, *47*, 809–813. [CrossRef] [PubMed]
76. Wong, B.H.; Silver, D.L. Mfsd2a: A physiologically important lysolipid transporter in the brain and eye. *Adv. Exp. Med. Biol.* **2020**, *1276*, 223–234. [CrossRef] [PubMed]
77. Sanchez-Campillo, M.; Ruiz-Palacios, M.; Ruiz-Alcaraz, A.J.; Prieto-Sanchez, M.T.; Blanco-Carnero, J.E.; Zornoza, M.; Ruiz-Pastor, M.J.; Demmelmair, H.; Sanchez-Solis, M.; Koletzko, B.; et al. Child head circumference and placental MFSD2a Expression are associated to the level of MFSD2a in maternal blood during pregnancy. *Front. Endocrinol.* **2020**, *11*, 38. [CrossRef]
78. Thies, F.; Delachambre, M.C.; Bentejac, M.; Lagarde, M.; Lecerf, J. Unsaturated fatty acids esterified in 2-acyl-l-lysophosphatidylcholine bound to albumin are more efficiently taken up by the young rat brain than the unesterified form. *J. Neurochem.* **1992**, *59*, 1110–1116. [CrossRef] [PubMed]
79. Chmurzynska, A. The multigene family of fatty acid-binding proteins (FABPs): Function, structure and polymorphism. *J. Appl. Genet.* **2006**, *47*, 39–48. [CrossRef]
80. Owada, Y.; Yoshimoto, T.; Kondo, H. Spatio-temporally differential expression of genes for three members of fatty acid binding proteins in developing and mature rat brains. *J. Chem. Neuroanat.* **1996**, *12*, 113–122. [CrossRef]
81. Cataltepe, O.; Arikan, M.C.; Ghelfi, E.; Karaaslan, C.; Ozsurekci, Y.; Dresser, K.; Li, Y.; Smith, T.W.; Cataltepe, S. Fatty acid binding protein 4 is expressed in distinct endothelial and non-endothelial cell populations in glioblastoma. *Neuropathol. Appl. Neurobiol.* **2012**, *38*, 400–410. [CrossRef]
82. Feng, L.; Hatten, M.E.; Heintz, N. Brain lipid-binding protein (BLBP): A novel signaling system in the developing mammalian CNS. *Neuron* **1994**, *12*, 895–908. [CrossRef]
83. Liang, Y.; Bollen, A.W.; Aldape, K.D.; Gupta, N. Nuclear FABP7 immunoreactivity is preferentially expressed in infiltrative glioma and is associated with poor prognosis in EGFR-overexpressing glioblastoma. *BMC Cancer* **2006**, *6*, 97. [CrossRef] [PubMed]
84. Kaloshi, G.; Mokhtari, K.; Carpentier, C.; Taillibert, S.; Lejeune, J.; Marie, Y.; Delattre, J.Y.; Godbout, R.; Sanson, M. FABP7 expression in glioblastomas: Relation to prognosis, invasion and EGFR status. *J. Neurooncol.* **2007**, *84*, 245–248. [CrossRef] [PubMed]
85. De Rosa, A.; Pellegatta, S.; Rossi, M.; Tunici, P.; Magnoni, L.; Speranza, M.C.; Malusa, F.; Miragliotta, V.; Mori, E.; Finocchiaro, G.; et al. A radial glia gene marker, fatty acid binding protein 7 (FABP7), is involved in proliferation and invasion of glioblastoma cells. *PLoS ONE* **2012**, *7*, e52113. [CrossRef]
86. Liu, R.Z.; Mita, R.; Beaulieu, M.; Gao, Z.; Godbout, R. Fatty acid binding proteins in brain development and disease. *Int. J. Dev. Biol.* **2010**, *54*, 1229–1239. [CrossRef] [PubMed]

87. Bernlohr, D.A.; Coe, N.R.; Simpson, M.A.; Hertzel, A.V. Regulation of gene expression in adipose cells by polyunsaturated fatty acids. *Adv. Exp. Med. Biol.* **1997**, *422*, 145–156. [CrossRef]

88. Murphy, E.J.; Owada, Y.; Kitanaka, N.; Kondo, H.; Glatz, J.F. Brain arachidonic acid incorporation is decreased in heart fatty acid binding protein gene-ablated mice. *Biochemistry* **2005**, *44*, 6350–6360. [CrossRef]

89. Hanhoff, T.; Lucke, C.; Spener, F. Insights into binding of fatty acids by fatty acid binding proteins. *Mol. Cell Biochem.* **2002**, *239*, 45–54. [CrossRef]

90. Balendiran, G.K.; Schnutgen, F.; Scapin, G.; Borchers, T.; Xhong, N.; Lim, K.; Godbout, R.; Spener, F.; Sacchettini, J.C. Crystal structure and thermodynamic analysis of human brain fatty acid-binding protein. *J. Biol. Chem.* **2000**, *275*, 27045–27054. [CrossRef]

91. Owada, Y. Fatty acid binding protein: Localization and functional significance in the brain. *Tohoku J. Exp. Med.* **2008**, *214*, 213–220. [CrossRef]

92. Maximin, E.; Langelier, B.; Aioun, J.; Al-Gubory, K.H.; Bordat, C.; Lavialle, M.; Heberden, C. Fatty acid binding protein 7 and n-3 poly unsaturated fatty acid supply in early rat brain development. *Dev. Neurobiol.* **2016**, *76*, 287–297. [CrossRef]

93. Kawakita, E.; Hashimoto, M.; Shido, O. Docosahexaenoic acid promotes neurogenesis in vitro and in vivo. *Neuroscience* **2006**, *139*, 991–997. [CrossRef]

94. Sakayori, N.; Maekawa, M.; Numayama-Tsuruta, K.; Katura, T.; Moriya, T.; Osumi, N. Distinctive effects of arachidonic acid and docosahexaenoic acid on neural stem /progenitor cells. *Genes Cells* **2011**, *16*, 778–790. [CrossRef] [PubMed]

95. Cheon, M.S.; Kim, S.H.; Fountoulakis, M.; Lubec, G. Heart type fatty acid binding protein (H-FABP) is decreased in brains of patients with Down syndrome and Alzheimer's disease. *J. Neural. Transm. Suppl.* **2003**, 225–234. [CrossRef]

96. Sanchez-Font, M.F.; Bosch-Comas, A.; Gonzalez-Duarte, R.; Marfany, G. Overexpression of FABP7 in Down syndrome fetal brains is associated with PKNOX1 gene-dosage imbalance. *Nucleic Acids Res.* **2003**, *31*, 2769–2777. [CrossRef] [PubMed]

97. Watanabe, A.; Toyota, T.; Owada, Y.; Hayashi, T.; Iwayama, Y.; Matsumata, M.; Ishitsuka, Y.; Nakaya, A.; Maekawa, M.; Ohnishi, T.; et al. Fabp7 maps to a quantitative trait locus for a schizophrenia endophenotype. *PLoS Biol.* **2007**, *5*, e297. [CrossRef] [PubMed]

98. Braff, D.L.; Geyer, M.A.; Light, G.A.; Sprock, J.; Perry, W.; Cadenhead, K.S.; Swerdlow, N.R. Impact of prepulse characteristics on the detection of sensorimotor gating deficits in schizophrenia. *Schizophr. Res.* **2001**, *49*, 171–178. [CrossRef]

99. Gearhart, D.A.; Toole, P.F.; Warren Beach, J. Identification of brain proteins that interact with 2-methylnorharman. An analog of the parkinsonian-inducing toxin, MPP+. *Neurosci. Res.* **2002**, *44*, 255–265. [CrossRef]

100. Steinacker, P.; Mollenhauer, B.; Bibl, M.; Cepek, L.; Esselmann, H.; Brechlin, P.; Lewczuk, P.; Poser, S.; Kretzschmar, H.A.; Wiltfang, J.; et al. Heart fatty acid binding protein as a potential diagnostic marker for neurodegenerative diseases. *Neurosci. Lett.* **2004**, *370*, 36–39. [CrossRef]

101. Mita, R.; Beaulieu, M.J.; Field, C.; Godbout, R. Brain fatty acid-binding protein and ω-3/ω-6 fatty acids. *J. Biol. Chem.* **2010**, *285*, 37005–37015. [CrossRef]

102. Cunnane, S.C.; Plourde, M.; Pifferi, F.; Begin, M.; Feart, C.; Barberger-Gateau, P. Fish, docosahexaenoic acid and Alzheimer's disease. *Prog. Lipid Res.* **2009**, *48*, 239–256. [CrossRef]

103. Calderon, F.; Kim, H.Y. Docosahexaenoic acid promotes neurite growth in hippocampal neurons. *J. Neurochem.* **2004**, *90*, 979–988. [CrossRef] [PubMed]

104. Cao, D.; Xue, R.; Xu, J.; Liu, Z. Effects of docosahexaenoic acid on the survival and neurite outgrowth of rat cortical neurons in primary cultures. *J. Nutr. Biochem.* **2005**, *16*, 538–546. [CrossRef] [PubMed]

105. Insua, M.F.; Garelli, A.; Rotstein, N.P.; German, O.L.; Arias, A.; Politi, L.E. Cell cycle regulation in retinal progenitors by glia-derived neurotrophic factor and docosahexaenoic acid. *Investig. Ophthalmol. Visual Sci.* **2003**, *44*, 2235–2244. [CrossRef] [PubMed]

106. Aïd, S.; Vancassel, S.; Poumès-Ballihaut, C.; Chalon, S.; Guesnet, P.; Lavialle, M. Effect of a diet-induced n-3 PUFA depletion on cholinergic parameters in the rat hippocampus. *J. Lipid Res.* **2003**, *44*, 1545–1551. [CrossRef] [PubMed]

107. Chalon, S. Omega-3 fatty acids and monoamine neurotransmission. *Prostaglandins Leukot. Essential Fatty Acids* **2006**, *75*, 259–269. [CrossRef] [PubMed]

108. Anderson, G.J.; Neuringer, M.; Lin, D.S.; Connor, W.E. Can prenatal N-3 fatty acid deficiency be completely reversed after birth? Effects on retinal and brain biochemistry and visual function in rhesus monkeys. *Pediatr. Res.* **2005**, *58*, 865–872. [CrossRef]

109. Innis, S.M. Dietary (n-3) Fatty Acids and Brain Development. *J. Nutr.* **2018**, *137*, 855–859. [CrossRef]

110. Coti Bertrand, P.; O'Kusky, J.R.; Innis, S.M. Maternal dietary (n-3) fatty acid deficiency alters neurogenesis in the embryonic rat brain. *J. Nutr.* **2006**, *136*, 1570–1575. [CrossRef]

111. Salem, N.; Litman, B.; Kim, H.Y.; Gawrisch, K. Mechanisms of action of docosahexaenoic acid in the nervous system. *Lipids* **2001**, *36*, 945–959. [CrossRef]

112. Lengqvist, J.; Mata de Urquiza, A.; Bergman, A.-C.; Willson, T.M.; Sjövall, J.; Perlmann, T.; Griffiths, W.J. Polyunsaturated fatty acids including docosahexaenoic and arachidonic acid bind to the retinoid X receptor α ligand-binding domain. *Mol. Cell Proteom.* **2004**, *3*, 692–703. [CrossRef]

113. Rioux, L.; Arnold, S.E. The expression of retinoic acid receptor alpha is increased in the granule cells of the dentate gyrus in schizophrenia. *Psychiatry Res.* **2005**, *133*, 13–21. [CrossRef]

114. Wada, K.; Nakajima, A.; Katayama, K.; Kudo, C.; Shibuya, A.; Kubota, N.; Terauchi, Y.; Tachibana, M.; Miyoshi, H.; Kamisaki, Y.; et al. Peroxisome proliferator-activated receptor γ-mediated regulation of neural stem cell proliferation and differentiation. *J. Biol. Chem.* **2006**, *281*, 12673–12681. [CrossRef]
115. Cao, D.H.; Xu, J.F.; Xue, R.H.; Zheng, W.F.; Liu, Z.L. Protective effect of chronic ethyl docosahexaenoate administration on brain injury in ischemic gerbils. *Pharmacol. Biochem. Behav.* **2004**, *79*, 651–659. [CrossRef] [PubMed]
116. Green, P.; Glozman, S.; Weiner, L.; Yavin, E. Enhanced free radical scavenging and decreased lipid peroxidation in the rat fetal brain after treatment with ethyl docosahexaenoate. *Biochim. Biophys. Acta Mol. Cell Biol. Lipids* **2001**, *1532*, 203–212. [CrossRef]
117. Okada, M.; Amamoto, T.; Tomonaga, M.; Kawachi, A.; Yazawa, K.; Mine, K.; Fujiwara, M. The chronic administration of docosahexaenoic acid reduces the spatial cognitive deficit following transient forebrain ischemia in rats. *Neuroscience* **1996**, *71*, 17–25. [CrossRef]
118. Samson, F.P.; Fabunmi, T.E.; Patrick, A.T.; Jee, D.; Gutsaeva, D.R.; Jahng, W.J. Fatty Acid Composition and Stoichiometry Determine the Angiogenesis Microenvironment. *ACS Omega* **2021**, *6*, 5953–5961. [CrossRef] [PubMed]
119. Wang, W.; Shinto, L.; Connor, W.E.; Quinn, J.F. Nutritional biomarkers in Alzheimer's disease: The association between carotenoids, n-3 fatty acids, and dementia severity. *J. Alzheimers Dis.* **2008**, *13*, 31–38. [CrossRef]
120. Cao, L.; Tan, L.; Wang, H.F.; Jiang, T.; Zhu, X.C.; Lu, H.; Tan, M.S.; Yu, J.T. Dietary patterns and risk of dementia: A systematic review and meta-analysis of cohort studies. *Mol. Neurobiol.* **2016**, *53*, 6144–6154. [CrossRef]
121. Bousquet, M.; Saint-Pierre, M.; Julien, C.; Salem, N., Jr.; Cicchetti, F.; Calon, F. Beneficial effects of dietary omega-3 polyunsaturated fatty acid on toxin-induced neuronal degeneration in an animal model of Parkinson's disease. *FASEB J.* **2008**, *22*, 1213–1225. [CrossRef] [PubMed]
122. Sinn, N.; Milte, C.M.; Street, S.J.; Buckley, J.D.; Coates, A.M.; Petkov, J.; Howe, P.R. Effects of n-3 fatty acids, EPA v. DHA, on depressive symptoms, quality of life, memory and executive function in older adults with mild cognitive impairment: A 6-month randomised controlled trial. *Br. J. Nutr.* **2012**, *107*, 1682–1693. [CrossRef]
123. Freund-Levi, Y.; Eriksdotter-Jonhagen, M.; Cederholm, T.; Basun, H.; Faxen-Irving, G.; Garlind, A.; Vedin, I.; Vessby, B.; Wahlund, L.O.; Palmblad, J. Omega-3 fatty acid treatment in 174 patients with mild to moderate Alzheimer disease: OmegAD study: A randomized double-blind trial. *Arch. Neurol.* **2006**, *63*, 1402–1408. [CrossRef] [PubMed]
124. Quinn, J.F.; Raman, R.; Thomas, R.G.; Yurko-Mauro, K.; Nelson, E.B.; Van Dyck, C.; Galvin, J.E.; Emond, J.; Jack, C.R., Jr.; Weiner, M.; et al. Docosahexaenoic acid supplementation and cognitive decline in Alzheimer disease: A randomized trial. *JAMA* **2010**, *304*, 1903–1911. [CrossRef]
125. Eilander, A.; Hundscheid, D.C.; Osendarp, S.J.; Transler, C.; Zock, P.L. Effects of n-3 long chain polyunsaturated fatty acid supplementation on visual and cognitive development throughout childhood: A review of human studies. *Prostaglandins Leukot. Essent. Fatty Acids* **2007**, *76*, 189–203. [CrossRef]
126. Christie, W.W.; Harwood, J.L. Oxidation of polyunsaturated fatty acids to produce lipid mediators. *Essays Biochem.* **2020**, *64*, 401–421. [CrossRef]
127. Oh, D.Y.; Walenta, E.; Akiyama, T.E.; Lagakos, W.S.; Lackey, D.; Pessentheiner, A.R.; Sasik, R.; Hah, N.; Chi, T.J.; Cox, J.M.; et al. A Gpr120-selective agonist improves insulin resistance and chronic inflammation in obese mice. *Nat. Med.* **2014**, *20*, 942–947. [CrossRef]
128. Diep, Q.N.; Touyz, R.M.; Schiffrin, E.L. Docosahexaenoic acid, a peroxisome proliferator-activated receptor-alpha ligand, induces apoptosis in vascular smooth muscle cells by stimulation of p38 mitogen-activated protein kinase. *Hypertension* **2000**, *36*, 851–855. [CrossRef]
129. Chitre, N.M.; Wood, B.J.; Ray, A.; Moniri, N.H.; Murnane, K.S. Docosahexaenoic acid protects motor function and increases dopamine synthesis in a rat model of Parkinson's disease via mechanisms associated with increased protein kinase activity in the striatum. *Neuropharmacology* **2020**, *167*, 107976. [CrossRef]
130. Zúñiga, J.; Cancino, M.; Medina, F.; Varela, P.; Vargas, R.; Tapia, G.; Videla, L.A.; Fernández, V. N-3 PUFA supplementation triggers PPAR-α activation and PPAR-α/NF-κB interaction: Anti-inflammatory implications in liver ischemia-reperfusion injury. *PLoS ONE* **2011**, *6*, e28502. [CrossRef]
131. Anderson, E.; Taylor, D. Stressing the heart of the matter: Re-thinking the mechanisms underlying therapeutic effects of n-3 polyunsaturated fatty acids. *F1000 Med. Rep.* **2012**, *4*, 13. [CrossRef]
132. Robertson, R.; Guihéneuf, F.; Schmid, M.; Stengel, D.; Fitzgerald, G.; Ross, P.; Stanton, C. *Algae-Derived Polyunsaturated Fatty Acids: Implications for Human Health*; Nova Sciences Publishers, Inc.: New York, NY, USA, 2013; p. 43.
133. Westphal, C.; Konkel, A.; Schunck, W.H. CYP-eicosanoids–A new link between omega-3 fatty acids and cardiac disease? *Prostaglandins Other Lipid Mediat.* **2011**, *96*, 99–108. [CrossRef]
134. Kuda, O. Bioactive metabolites of docosahexaenoic acid. *Biochimie* **2017**, *136*, 12–20. [CrossRef]
135. Serhan, C.N. Pro-resolving lipid mediators are leads for resolution physiology. *Nature* **2014**, *510*, 92–101. [CrossRef]
136. Bisicchia, E.; Sasso, V.; Catanzaro, G.; Leuti, A.; Besharat, Z.M.; Chiacchiarini, M.; Molinari, M.; Ferretti, E.; Viscomi, M.T.; Chiurchiù, V. Resolvin D1 halts remote neuroinflammation and improves functional recovery after focal brain damage via ALX/FPR2 receptor-regulated microRNAs. *Mol. Neurobiol.* **2018**, *55*, 6894–6905. [CrossRef]
137. Recchiuti, A.; Krishnamoorthy, S.; Fredman, G.; Chiang, N.; Serhan, C.N. MicroRNAs in resolution of acute inflammation: Identification of novel resolvin D1-miRNA circuits. *FASEB J.* **2010**, *25*, 544–560. [CrossRef]

138. Zuo, G.; Zhang, D.; Mu, R.; Shen, H.; Li, X.; Wang, Z.; Li, H.; Chen, G. Resolvin D2 protects against cerebral ischemia/reperfusion injury in rats. *Mol. Brain* **2018**, *11*, 1–13. [CrossRef]
139. Dalli, J.; Winkler, J.W.; Colas, R.A.; Arnardottir, H.; Cheng, C.Y.C.; Chiang, N.; Petasis, N.A.; Serhan, C.N. Resolvin D3 and aspirin-triggered resolvin D3 are potent immunoresolvents. *Chem. Biol.* **2013**, *20*, 188–201. [CrossRef]
140. Hong, S.; Tjonahen, E.; Morgan, E.L.; Lu, Y.; Serhan, C.N.; Rowley, A.F. Rainbow trout (*Oncorhynchus mykiss*) brain cells biosynthesize novel docosahexaenoic acid-derived resolvins and protectins-Mediator lipidomic analysis. *Prostaglandins Other Lipid Mediat.* **2005**, *78*, 107–116. [CrossRef]
141. Ariel, A.; Serhan, C.N. New lives given by cell death: Macrophage differentiation following their encounter with apoptotic leukocytes during the resolution of inflammation. *Front. Immunol.* **2012**, *3*, 4. [CrossRef]
142. Stables, M.J.; Shah, S.; Camon, E.B.; Lovering, R.C.; Newson, J.; Bystrom, J.; Farrow, S.; Gilroy, D.W. Transcriptomic analyses of murine resolution-phase macrophages. *Blood* **2011**, *118*, 192–208. [CrossRef]
143. Zhang, M.J.; Spite, M. Resolvins: Anti-inflammatory and proresolving mediators derived from omega-3 polyunsaturated fatty acids. *Ann. Rev. Nutr.* **2012**, *32*, 203–207. [CrossRef]
144. Serhan, C.N.; Dalli, J.; Colas, R.A.; Winkler, J.W.; Chiang, N. Protectins and maresins: New pro-resolving families of mediators in acute inflammation and resolution bioactive metabolome. *Biochim. Biophys. Acta* **2015**, *1851*, 397–413. [CrossRef]
145. Serhan, C.N.; Dalli, J.; Karamnov, S.; Choi, A.; Park, C.-K.; Xu, Z.-Z.; Ji, R.-R.; Zhu, M.; Petasis, N.A. Macrophage proresolving mediator maresin 1 stimulates tissue regeneration and controls pain. *FASEB J.* **2012**, *26*, 1755–1765. [CrossRef]
146. Serhan, C.N.; Yang, R.; Martinod, K.; Kasuga, K.; Pillai, P.S.; Porter, T.F.; Oh, S.F.; Spite, M. Maresins: Novel macrophage mediators with potent antiinflammatory and proresolving actions. *J. Exp. Med.* **2008**, *206*, 15–23. [CrossRef]
147. Yanes, O.; Clark, J.; Wong, D.M.; Patti, G.J.; Sánchez-Ruiz, A.; Benton, H.P.; Trauger, S.A.; Desponts, C.; Ding, S.; Siuzdak, G. Metabolic oxidation regulates embryonic stem cell differentiation. *Nat. Chem. Biol.* **2010**, *6*, 411–417. [CrossRef]
148. Yang, T.; Xu, G.; Newton, P.T.; Chagin, A.S.; Mkrtchian, S.; Carlström, M.; Zhang, X.M.; Harris, R.A.; Cooter, M.; Berger, M.; et al. Maresin 1 attenuates neuroinflammation in a mouse model of perioperative neurocognitive disorders. *Br. J. Anaesth.* **2019**, *122*, 350–360. [CrossRef]
149. Francos-Quijorna, I.; Santos-Nogueira, E.; Gronert, K.; Sullivan, A.B.; Kopp, M.A.; Brommer, B.; David, S.; Schwab, J.M.; López-Vales, R. Maresin 1 promotes inflammatory resolution, neuroprotection, and functional neurological recovery after spinal cord injury. *J. Neurosci.* **2017**, *37*, 11731–11743. [CrossRef]
150. Chiang, N.; Libreros, S.; Norris, P.C.; de la Rosa, X.; Serhan, C.N. Maresin 1 activates LGR6 receptor promoting phagocyte immunoresolvent functions. *J. Clin. Investig.* **2019**, *129*, 5294–5311. [CrossRef]
151. Bazan, N.G.; Molina, M.F.; Gordon, W.C. Docosahexaenoic acid signalolipidomics in nutrition: Significance in aging, neuroinflammation, macular degeneration, Alzheimer's, and other neurodegenerative diseases. *Ann. Rev. Nutr.* **2011**, *31*, 321–351. [CrossRef]
152. Balas, L.; Durand, T. Dihydroxylated E, E, Z-docosatrienes. An overview of their synthesis and biological significance. *Prog. Lipid Res.* **2016**, *61*, 1–18. [CrossRef]
153. Bazan, N.G. Neuroprotectin D1 (NPD1): A DHA-derived mediator that protects brain and retina against cell injury-induced oxidative stress. *Brain Pathol.* **2005**, *15*, 15267–15278. [CrossRef]
154. Calandria, J.M.; Marcheselli, V.L.; Mukherjee, P.K.; Uddin, J.; Winkler, J.W.; Petasis, N.A.; Bazan, N.G. Selective survival rescue in 15-lipoxygenase-1-deficient retinal pigment epithelial cells by the novel docosahexaenoic acid-derived mediator, neuroprotectin D1. *J. Biol. Chem.* **2009**, *284*, 17877–17882. [CrossRef]
155. Morita, M.; Kuba, K.; Ichikawa, A.; Nakayama, M.; Katahira, J.; Iwamoto, R.; Watanebe, T.; Sakabe, S.; Daidoji, T.; Nakamura, S.; et al. The lipid mediator protectin D1 inhibits influenza virus replication and improves severe influenza. *Cell* **2013**, *153*, 112–125. [CrossRef]
156. Zhao, Y.; Calon, F.; Julien, C.; Winkler, J.W.; Petasis, N.A.; Lukiw, W.J.; Bazan, N.G. Docosahexaenoic acid-derived neuroprotectin D1 induces neuronal survival via secretase- and PPARγ-mediated mechanisms in Alzheimer's disease models. *PLoS ONE* **2011**, *6*, e15816. [CrossRef] [PubMed]
157. Garcia-Sastre, A. Lessons from lipids in the fight against influenza. *Cell* **2013**, *154*, 22–23. [CrossRef] [PubMed]
158. Gladine, C.; Laurie, J.-C.; Giulia, C.; Dominique, B.; Corinne, C.; Nathalie, H.; Bart, S.; Giuseppe, Z.; Alessio, P.; Jean-Marie, G.; et al. Neuroprostanes, produced by free-radical mediated peroxidation of DHA, inhibit the inflammatory response of human macrophages. *Free Radic. Biol. Med.* **2014**, *75*, S15–S25. [CrossRef] [PubMed]
159. Groeger, A.L.; Cipollina, C.; Cole, M.P.; Woodcock, S.R.; Bonacci, G.; Rudolph, T.K.; Rudolph, V.; Freeman, B.A.; Schopfer, F.J. Cyclooxygenase-2 generates anti-inflammatory mediators from omega-3 fatty acids. *Nat. Chem. Biol.* **2010**, *6*, 433–441. [CrossRef] [PubMed]
160. Dyall, S.C. Long-chain omega-3 fatty acids and the brain: A review of the independent and shared effects of EPA, DPA and DHA. *Front Aging Neurosci.* **2015**, *7*, 52. [CrossRef]
161. De Petrocellis, L.; Melck, D.; Bisogno, T.; Milone, A.; Di Marzo, V. Finding of the endocannabinoid signalling system in Hydra, a very primitive organism: Possible role in the feeding response. *Neuroscience* **1999**, *92*, 377–387. [CrossRef]
162. Kim, J.; Carlson, M.E.; Watkins, B.A. Docosahexaenoyl ethanolamide improves glucose uptake and alters endocannabinoid system gene expression in proliferating and differentiating C2C12 myoblasts. *Front. Physiol.* **2014**, *5*. [CrossRef]

163. Soderstrom, K. Endocannabinoids link feeding state and auditory perception-related gene expression. *J. Neurosci.* **2004**, *24*, 10013–10024. [CrossRef]

164. Valenti, M.; Cottone, E.; Martinez, R.; De Pedro, N.; Rubio, M.; Viveros, M.P.; Franzoni, M.F.; Delgado, M.J.; Di Marzo, V. The endocannabinoid system in the brain of Carassius auratus and its possible role in the control of food intake. *J. Neurochem.* **2005**, *95*, 662–672. [CrossRef]

165. Chiang, N.; Takano, T.; Arita, M.; Watanabe, S.; Serhan, C.N. A novel rat lipoxin A4 receptor that is conserved in structure and function. *Br. J. Pharmacol.* **2003**, *139*, 89–98. [CrossRef]

166. Piazza, P.V.; Lafontan, M.; Girard, J. Integrated physiology and pathophysiology of CB1-mediated effects of the endocannabinoid system. *Diabetes Metab.* **2007**, *33*, 97–107. [CrossRef]

167. Masoodi, M.; Kuda, O.; Rossmeisl, M.; Flachs, P.; Kopecky, J. Lipid signaling in adipose tissue: Connecting inflammation & metabolism. *Biochim. Biophys. Acta Mol. Cell Biol. Lipids* **2015**, *1851*, 503–518.

168. D'Addario, C.; Micioni Di Bonaventura, M.V.; Pucci, M.; Romano, A.; Gaetani, S.; Ciccocioppo, R.; Cifani, C.; Maccarrone, M. Endocannabinoid signaling and food addiction. *Neurosci. Biobehav. Rev.* **2014**, *47*, 203–224. [CrossRef] [PubMed]

169. Wagner, J.J.; Alger, B.E. Increased neuronal excitability during depolarization-induced suppression of inhibition in rat hippocampus. *J. Physiol.* **1996**, *495*, 107–112. [CrossRef]

170. Wilson, R.I.; Nicoll, R.A. Neuroscience: Endocannabinoid signaling in the brain. *Science* **2002**, *296*, 678–682. [CrossRef] [PubMed]

171. Calder, P.C. Omega-3 polyunsaturated fatty acids and inflammatory processes: Nutrition or pharmacology? *Br. J. Clin. Pharmacol.* **2013**, *75*, 645–662. [CrossRef] [PubMed]

172. Lu, H.C.; MacKie, K. An introduction to the endogenous cannabinoid system. *Biol. Psychiatry* **2016**. [CrossRef]

173. Song, C.; Manku, M.S.; Horrobin, D.F. Long-chain polyunsaturated fatty acids modulate interleukin-1β–induced changes in behavior, monoaminergic neurotransmitters, and brain inflammation in rats. *J. Nutr.* **2018**. [CrossRef] [PubMed]

174. Bozzatello, P.; Brignolo, E.; De Grandi, E.; Bellino, S. Supplementation with omega-3 fatty acids in psychiatric disorders: A review of literature data. *J. Clin. Med.* **2016**, *5*, 67. [CrossRef]

175. Hibbeln, J.R. Fish consumption and major depression. *Lancet* **1998**, *351*, 1213. [CrossRef]

176. Kiliaan, A.; Königs, A. Critical appraisal of omega-3 fatty acids in attention-deficit/hyperactivity disorder treatment. *Neuropsychiatr. Dis. Treat.* **2016**, *12*, 1869–1882. [CrossRef] [PubMed]

177. Lukiw, W.J.; Cui, J.G.; Marcheselli, V.L.; Bodker, M.; Botkjaer, A.; Gotlinger, K.; Serhan, C.N.; Bazan, N.G. A role for docosahexaenoic acid-derived neuroprotectin D1 in neural cell survival and Alzheimer disease. *J. Clin. Investig.* **2005**, *115*, 2774–2783. [CrossRef]

178. Boudrault, C.; Bazinet, R.P.; Ma, D.W.L. Experimental models and mechanisms underlying the protective effects of n-3 polyunsaturated fatty acids in Alzheimer's disease. *J. Nutr. Biochem.* **2009**, *20*, 1–10. [CrossRef]

179. Hashimoto, M.; Hossain, S.; Agdul, H.; Shido, O. Docosahexaenoic acid-induced amelioration on impairment of memory learning in amyloid β-infused rats relates to the decreases of amyloid β and cholesterol levels in detergent-insoluble membrane fractions. *Biochim. Biophys. Acta Mol. Cell Biol. Lipids* **2005**, *1738*, 91–98. [CrossRef]

180. Seidl, S.E.; Santiago, J.A.; Bilyk, H.; Potashkin, J.A. The emerging role of nutrition in Parkinson's disease. *Front. Aging Neurosci.* **2014**, *6*. [CrossRef] [PubMed]

181. McNamara, R.K. DHA Deficiency and prefrontal cortex neuropathology in recurrent affective disorders. *J. Nutr.* **2010**, *140*, 864–868. [CrossRef] [PubMed]

182. McNamara, R.K.; Hahn, C.G.; Jandacek, R.; Rider, T.; Tso, P.; Stanford, K.E.; Richtand, N.M. Selective deficits in the omega-3 fatty acid docosahexaenoic acid in the postmortem orbitofrontal cortex of patients with major depressive disorder. *Biol. Psychiatry* **2007**, *62*, 17–24. [CrossRef] [PubMed]

183. Basak, S.; Vilasagaram, S.; Duttaroy, A.K. Maternal dietary deficiency of n-3 fatty acids affects metabolic and epigenetic phenotypes of the developing fetus. *Prostaglandins Leukot. Essent. Fatty Acids* **2020**, *158*, 102109. [CrossRef]

184. Srinivas, V.; Molangiri, A.; Mallepogu, A.; Kona, S.R.; Ibrahim, A.; Duttaroy, A.K.; Basak, S. Maternal n-3 PUFA deficiency alters uterine artery remodeling and placental epigenome in the mice. *J. Nutr. Biochem.* **2021**, in press. [CrossRef]

185. Boucher, O.; Burden, M.J.; Muckle, G.; Saint-Amour, D.; Ayotte, P.; Dewailly, E.; Nelson, C.A.; Jacobson, S.W.; Jacobson, J.L. Neurophysiologic and neurobehavioral evidence of beneficial effects of prenatal omega-3 fatty acid intake on memory function at school age. *Am. J. Clin. Nutr.* **2011**, *93*, 1025–1037. [CrossRef]

186. Mulder, K.A.; King, D.J.; Innis, S.M. Omega-3 fatty acid deficiency in infants before birth identified using a randomized trial of maternal DHA supplementation in pregnancy. *PLoS ONE* **2014**, *9*, e83764. [CrossRef]

187. Bos, D.J.; Oranje, B.; Veerhoek, E.S.; Van Diepen, R.M.; Weusten, J.M.H.; Demmelmair, H.; Koletzko, B.; De Sain-Van Der Velden, M.G.M.; Eilander, A.; Hoeksma, M.; et al. Reduced symptoms of inattention after dietary omega-3 fatty acid supplementation in boys with and without attention deficit/hyperactivity disorder. *Neuropsychopharmacology* **2015**, *40*, 2298–2306. [CrossRef]

188. Parellada, M.; Llorente, C.; Calvo, R.; Gutierrez, S.; Lázaro, L.; Graell, M.; Guisasola, M.; Dorado, M.L.; Boada, L.; Romo, J.; et al. Randomized trial of omega-3 for autism spectrum disorders: Effect on cell membrane composition and behavior. *Eur. Neuropsychopharmacol.* **2017**, *27*, 1319–1330. [CrossRef] [PubMed]

189. Bauer, I.; Hughes, M.; Rowsell, R.; Cockerell, R.; Pipingas, A.; Crewther, S.; Crewther, D. Omega-3 supplementation improves cognition and modifies brain activation in young adults. *Hum. Psychopharmacol.* **2014**, *29*, 133–144. [CrossRef] [PubMed]

190. Jaremka, L.M.; Derry, H.M.; Bornstein, R.; Prakash, R.S.; Peng, J.; Belury, M.A.; Andridge, R.R.; Malarkey, W.B.; Kiecolt-Glaser, J.K. Omega-3 supplementation and loneliness-related memory problems: Secondary analyses of a randomized controlled trial. *Psychosom. Med.* **2014**, *76*, 650–658. [CrossRef]

191. Smith, S.L.; Rouse, C.A. Docosahexaenoic acid and the preterm infant. *Mater. Health Neonatol. Perinatol.* **2017**, *3*, 22. [CrossRef]

192. Albanese, E.; Dangour, A.D.; Uauy, R.; Acosta, D.; Guerra, M.; Guerra, S.S.; Huang, Y.; Jacob, K.S.; de Rodriguez, J.L.; Noriega, L.H.; et al. Dietary fish and meat intake and dementia in Latin America, China, and India: A 10/66 Dementia Research Group population-based study. *Am. J. Clin. Nutr.* **2009**, *90*, 392–400. [CrossRef] [PubMed]

193. Milte, C.M.; Sinn, N.; Street, S.J.; Buckley, J.D.; Coates, A.M.; Howe, P.R. Erythrocyte polyunsaturated fatty acid status, memory, cognition and mood in older adults with mild cognitive impairment and healthy controls. *Prostaglandins Leukot. Essent. Fatty Acids* **2011**, *84*, 153–161. [CrossRef]

194. Yin, Y.; Fan, Y.; Lin, F.; Xu, Y.; Zhang, J. Nutrient biomarkers and vascular risk factors in subtypes of mild cognitive impairment: A cross-sectional study. *J. Nutr. Health Aging.* **2015**, *19*, 39–47. [CrossRef]

195. Bazinet, R.P.; Layé, S. Polyunsaturated fatty acids and their metabolites in brain function and disease. *Nat. Rev. Neurosci.* **2014**, *15*, 771–785. [CrossRef]

196. Shamim, A.; Mahmood, T.; Ahsan, F.; Kumar, A.; Bagga, P. Lipids: An insight into the neurodegenerative disorders. *Clin. Nutr. Exp.* **2018**, *20*, 1–19. [CrossRef]

197. Lopez, L.B.; Kritz-Silverstein, D.; Barrett Connor, E. High dietary and plasma levels of the omega-3 fatty acid docosahexaenoic acid are associated with decreased dementia risk: The Rancho Bernardo study. *J. Nutr. Health Aging.* **2011**, *15*, 25–31. [CrossRef]

198. Phillips, M.A.; Childs, C.E.; Calder, P.C.; Rogers, P.J. Lower omega-3 fatty acid intake and status are associated with poorer cognitive function in older age: A comparison of individuals with and without cognitive impairment and Alzheimer's disease. *Nutr. Neurosci.* **2012**, *15*, 271–277. [CrossRef] [PubMed]

199. Schaefer, E.J.; Bongard, V.; Beiser, A.S.; Lamon-Fava, S.; Robins, S.J.; Au, R.; Tucker, K.L.; Kyle, D.J.; Wilson, P.W.; Wolf, P.A. Plasma phosphatidylcholine docosahexaenoic acid content and risk of dementia and Alzheimer disease: The Framingham Heart Study. *Arch. Neurol.* **2006**, *63*, 1545–1550. [CrossRef] [PubMed]

200. Shatenstein, B.; Kergoat, M.J.; Reid, I. Poor nutrient intakes during 1-year follow-up with community-dwelling older adults with early-stage Alzheimer dementia compared to cognitively intact matched controls. *J. Am. Diet Assoc.* **2007**, *107*, 2091–2099. [CrossRef]

201. Kotani, S.; Sakaguchi, E.; Warashina, S.; Matsukawa, N.; Ishikura, Y.; Kiso, Y.; Sakakibara, M.; Yoshimoto, T.; Guo, J.; Yamashima, T. Dietary supplementation of arachidonic and docosahexaenoic acids improves cognitive dysfunction. *Neurosci. Res.* **2006**, *56*, 159–164. [CrossRef] [PubMed]

202. Wurtman, R.J.; Ulus, I.H.; Cansev, M.; Watkins, C.J.; Wang, L.; Marzloff, G. Synaptic proteins and phospholipids are increased in gerbil brain by administering uridine plus docosahexaenoic acid orally. *Brain Res.* **2006**, *1088*, 83–92. [CrossRef]

203. Gomez-Soler, M.; Cordobilla, B.; Morato, X.; Fernandez-Duenas, V.; Domingo, J.C.; Ciruela, F. Triglyceride Form of Docosahexaenoic Acid Mediates Neuroprotection in Experimental Parkinsonism. *Front. Neurosci.* **2018**, *12*, 604. [CrossRef] [PubMed]

204. Patrick, R.P.; Ames, B.N. Vitamin D and the omega-3 fatty acids control serotonin synthesis and action, part 2: Relevance for ADHD, bipolar disorder, schizophrenia, and impulsive behavior. *FASEB J.* **2015**, *29*, 2207–2222. [CrossRef]

205. McNamara, R.K.; Jandacek, R.; Rider, T.; Tso, P.; Hahn, C.G.; Richtand, N.M.; Stanford, K.E. Abnormalities in the fatty acid composition of the postmortem orbitofrontal cortex of schizophrenic patients: Gender differences and partial normalization with antipsychotic medications. *Schizophr. Res.* **2007**, *91*, 37–50. [CrossRef]

206. Sinn, N.; Bryan, J. Effect of supplementation with polyunsaturated fatty acids and micronutrients on learning and behavior problems associated with child ADHD. *J. Dev. Behav. Pediatr.* **2007**, *28*, 82–91. [CrossRef]

207. Sinn, N.; Bryan, J.; Wilson, C. Cognitive effects of polyunsaturated fatty acids in children with attention deficit hyperactivity disorder symptoms: A randomised controlled trial. *Prostaglandins Leukot. Essent. Fatty Acids* **2008**, *78*, 311–326. [CrossRef] [PubMed]

208. Belanger, S.A.; Vanasse, M.; Spahis, S.; Sylvestre, M.P.; Lippe, S.; L'Heureux, F.; Ghadirian, P.; Vanasse, C.M.; Levy, E. Omega-3 fatty acid treatment of children with attention-deficit hyperactivity disorder: A randomized, double-blind, placebo-controlled study. *Paediatr. Child Health* **2009**, *14*, 89–98. [CrossRef]

209. Johnson, M.; Ostlund, S.; Fransson, G.; Kadesjo, B.; Gillberg, C. Omega-3/omega-6 fatty acids for attention deficit hyperactivity disorder: A randomized placebo-controlled trial in children and adolescents. *J. Atten. Disord.* **2009**, *12*, 394–401. [CrossRef] [PubMed]

210. Gustafsson, P.A.; Birberg-Thornberg, U.; Duchen, K.; Landgren, M.; Malmberg, K.; Pelling, H.; Strandvik, B.; Karlsson, T. EPA supplementation improves teacher-rated behaviour and oppositional symptoms in children with ADHD. *Acta Paediatr.* **2010**, *99*, 1540–1549. [CrossRef]

211. Hariri, M.; Djazayery, A.; Djalali, M.; Saedisomeolia, A.; Rahimi, A.; Abdolahian, E. Effect of n-3 supplementation on hyperactivity, oxidative stress and inflammatory mediators in children with attention-deficit-hyperactivity disorder. *Malays J. Nutr.* **2012**, *18*, 329–335.

212. Behdani, F.; Hebrani, P.; Naseraee, A.; Haghighi, M.B.; Akhavanrezayat, A. Does omega-3 supplement enhance the therapeutic results of methylphenidate in attention deficit hyperactivity disorder patients? *J. Res. Med. Sci.* **2013**, *18*, 653–658. [PubMed]

213. Dashti, N.; Hekmat, H.; Soltani, H.R.; Rahimdel, A.; Javaherchian, M. Comparison of therapeutic effects of omega-3 and methylphenidate (ritalin((R))) in treating children with attention deficit hyperactivity disorder. *Iran J. Psychiatry Behav. Sci.* **2014**, *8*, 7–11.

214. Anand, P.; Sachdeva, A. Effect of poly unsaturated fatty acids administration on children with attention deficit hyperactivity disorder: A randomized controlled trial. *J. Clin. Diagn. Res.* **2016**, *10*, OC01–OC05. [CrossRef] [PubMed]

215. Assareh, M.; Davari Ashtiani, R.; Khademi, M.; Jazayeri, S.; Rai, A.; Nikoo, M. Efficacy of Polyunsaturated Fatty Acids (PUFA) in the treatment of attention deficit hyperactivity disorder. *J. Atten. Disord.* **2017**, *21*, 78–85. [CrossRef]

216. Kean, J.D.; Sarris, J.; Scholey, A.; Silberstein, R.; Downey, L.A.; Stough, C. Reduced inattention and hyperactivity and improved cognition after marine oil extract (PCSO-524(R)) supplementation in children and adolescents with clinical and subclinical symptoms of attention-deficit hyperactivity disorder (ADHD): A randomised, double-blind, placebo-controlled trial. *Psychopharmacology* **2017**, *234*, 403–420. [CrossRef] [PubMed]

217. Voigt, R.G.; Llorente, A.M.; Jensen, C.L.; Fraley, J.K.; Berretta, M.C.; Heird, W.C. A randomized, double-blind, placebo-controlled trial of docosahexaenoic acid supplementation in children with attention-deficit/hyperactivity disorder. *J. Pediatr.* **2001**, *139*, 189–196. [CrossRef] [PubMed]

218. Amminger, G.P.; Schafer, M.R.; Papageorgiou, K.; Klier, C.M.; Cotton, S.M.; Harrigan, S.M.; Mackinnon, A.; McGorry, P.D.; Berger, G.E. Long-chain omega-3 fatty acids for indicated prevention of psychotic disorders: A randomized, placebo-controlled trial. *Arch. Gen. Psychiatry* **2010**, *67*, 146–154. [CrossRef]

219. Pawelczyk, T.; Grancow-Grabka, M.; Kotlicka-Antczak, M.; Trafalska, E.; Pawelczyk, A. A randomized controlled study of the efficacy of six-month supplementation with concentrated fish oil rich in omega-3 polyunsaturated fatty acids in first episode schizophrenia. *J. Psychiatr Res.* **2016**, *73*, 34–44. [CrossRef]

220. Ammann, E.M.; Pottala, J.V.; Harris, W.S.; Espeland, M.A.; Wallace, R.; Denburg, N.L.; Carnahan, R.M.; Robinson, J.G. omega-3 fatty acids and domain-specific cognitive aging: Secondary analyses of data from WHISCA. *Neurology* **2013**, *81*, 1484–1491. [CrossRef]

221. Maekawa, M.; Watanabe, A.; Iwayama, Y.; Kimura, T.; Hamazaki, K.; Balan, S.; Ohba, H.; Hisano, Y.; Nozaki, Y.; Ohnishi, T.; et al. Polyunsaturated fatty acid deficiency during neurodevelopment in mice models the prodromal state of schizophrenia through epigenetic changes in nuclear receptor genes. *Transl. Psychiatry* **2017**, *7*, e1229. [CrossRef]

222. McNamara, R.K.; Jandacek, R.; Tso, P.; Blom, T.J.; Welge, J.A.; Strawn, J.R.; Adler, C.M.; Strakowski, S.M.; DelBello, M.P. Adolescents with or at ultra-high risk for bipolar disorder exhibit erythrocyte docosahexaenoic acid and eicosapentaenoic acid deficits: A candidate prodromal risk biomarker. *Early Interv. Psychiatry* **2016**, *10*, 203–211. [CrossRef]

223. Stoll, A.L.; Severus, W.E.; Freeman, M.P.; Rueter, S.; Zboyan, H.A.; Diamond, E.; Cress, K.K.; Marangell, L.B. Omega 3 fatty acids in bipolar disorder: A preliminary double-blind, placebo-controlled trial. *Arch. Gen. Psychiatry* **1999**, *56*, 407–412. [CrossRef] [PubMed]

224. Chiu, C.C.; Huang, S.Y.; Chen, C.C.; Su, K.P. Omega-3 fatty acids are more beneficial in the depressive phase than in the manic phase in patients with bipolar I disorder. *J. Clin. Psychiatry* **2005**, *66*, 1613–1614. [CrossRef] [PubMed]

225. Clayton, E.H.; Hanstock, T.L.; Hirneth, S.J.; Kable, C.J.; Garg, M.L.; Hazell, P.L. Reduced mania and depression in juvenile bipolar disorder associated with long-chain omega-3 polyunsaturated fatty acid supplementation. *Eur. J. Clin. Nutr.* **2009**, *63*, 1037–1040. [CrossRef] [PubMed]

226. Meldrum, S.J.; D'Vaz, N.; Simmer, K.; Dunstan, J.A.; Hird, K.; Prescott, S.L. Effects of high-dose fish oil supplementation during early infancy on neurodevelopment and language: A randomised controlled trial. *Br. J. Nutr.* **2012**, *108*, 1443–1454. [CrossRef]

227. Birch, E.E.; Garfield, S.; Castaneda, Y.; Hughbanks-Wheaton, D.; Uauy, R.; Hoffman, D. Visual acuity and cognitive outcomes at 4 years of age in a double-blind, randomized trial of long-chain polyunsaturated fatty acid-supplemented infant formula. *Early Hum. Dev.* **2007**, *83*, 279–284. [CrossRef]

228. Osendarp, S.J.; Baghurst, K.I.; Bryan, J.; Calvaresi, E.; Hughes, D.; Hussaini, M.; Karyadi, S.J.; van Klinken, B.J.; van der Knaap, H.C.; Lukito, W.; et al. Effect of a 12-mo micronutrient intervention on learning and memory in well-nourished and marginally nourished school-aged children: 2 parallel, randomized, placebo-controlled studies in Australia and Indonesia. *Am. J. Clin. Nutr.* **2007**, *86*, 1082–1093. [CrossRef]

229. Richardson, A.J.; Burton, J.R.; Sewell, R.P.; Spreckelsen, T.F.; Montgomery, P. Docosahexaenoic acid for reading, cognition and behavior in children aged 7-9 years: A randomized, controlled trial (the DOLAB Study). *PLoS ONE* **2012**, *7*, e43909. [CrossRef]

230. Dalton, A.; Wolmarans, P.; Witthuhn, R.C.; van Stuijvenberg, M.E.; Swanevelder, S.A.; Smuts, C.M. A randomised control trial in schoolchildren showed improvement in cognitive function after consuming a bread spread, containing fish flour from a marine source. *Prostaglandins Leukot. Essent. Fatty Acids* **2009**, *80*, 143–149. [CrossRef]

231. Brew, B.K.; Toelle, B.G.; Webb, K.L.; Almqvist, C.; Marks, G.B.; investigators, C. Omega-3 supplementation during the first 5 years of life and later academic performance: A randomised controlled trial. *Eur. J. Clin. Nutr.* **2015**, *69*, 419–424. [CrossRef]

232. Lassek, W.D.; Gaulin, S.J. Sex differences in the relationship of dietary Fatty acids to cognitive measures in american children. *Front. Evol. Neurosci.* **2011**, *3*, 5. [CrossRef]

233. Shulkin, M.; Pimpin, L.; Bellinger, D.; Kranz, S.; Fawzi, W.; Duggan, C.; Mozaffarian, D. N-3 fatty acid supplementation in mothers, preterm infants, and term infants and childhood psychomotor and visual development: A systematic review and meta-analysis. *J. Nutr.* **2018**, *149*, 409–418. [CrossRef] [PubMed]

234. Colombo, J.; Shaddy, D.J.; Gustafson, K.; Gajewski, B.J.; Thodosoff, J.M.; Kerling, E.; Carlson, S.E. The Kansas University DHA Outcomes Study (KUDOS) clinical trial: Long-term behavioral follow-up of the effects of prenatal DHA supplementation. *Am. J. Clin. Nutr.* **2019**, *109*, 1380–1392. [CrossRef]
235. Keim, S.A.; Boone, K.M.; Klebanoff, M.A.; Turner, A.N.; Rausch, J.; Nelin, M.A.; Rogers, L.K.; Yeates, K.O.; Nelin, L.; Sheppard, K.W. Effect of docosahexaenoic add supplementation vs placebo on developmental outcomes of toddlers born preterm a randomized clinical. *JAMA Pediatr.* **2018**, *172*, 1126–1134. [CrossRef]
236. Makrides, M.; Gibson, R.A.; McPhee, A.J.; Yelland, L.; Quinlivan, J.; Ryan, P.; Doyle, L.W.; Anderson, P.; Else, P.L.; Meyer, B.J.; et al. Effect of DHA supplementation during pregnancy on maternal depression and neurodevelopment of young children: A randomized controlled trial. *JAMA J. Am. Med. Assoc.* **2010**, *304*, 1675–1683. [CrossRef]
237. Collins, C.T.; Gibson, R.A.; Anderson, P.J.; McPhee, A.J.; Sullivan, T.R.; Gould, J.F.; Ryan, P.; Doyle, L.W.; Davis, P.G.; McMichael, J.E.; et al. Neurodevelopmental outcomes at 7 years' corrected age in preterm infants who were fed high-dose docosahexaenoic acid to term equivalent: A follow-up of a randomised controlled trial. *BMJ Open* **2015**, *5*. [CrossRef] [PubMed]
238. Smithers, L.G.; Collins, C.T.; Simmonds, L.A.; Gibson, R.A.; McPhee, A.; Makrides, M. Feeding preterm infants milk with a higher dose of docosahexaenoic acid than that used in current practice does not influence language or behavior in early childhood: A follow-up study of a randomized controlled trial. *Am. J. Clin. Nutr.* **2010**, *91*, 628–634. [CrossRef] [PubMed]
239. Basak, S.; Duttaroy, A.K. Effects of fatty acids on angiogenic activity in the placental extravillious trophoblast cells. *Prostaglandins Leukot. Essent. Fatty Acids* **2013**, *88*, 155–162. [CrossRef] [PubMed]
240. Bosetti, F. Arachidonic acid metabolism in brain physiology and pathology: Lessons from genetically altered mouse models. *J. Neurochem.* **2007**, *102*, 577–586. [CrossRef] [PubMed]
241. Harauma, A.; Hatanaka, E.; Yasuda, H.; Nakamura, M.T.; Salem, N., Jr.; Moriguchi, T. Effects of arachidonic acid, eicosapentaenoic acid and docosahexaenoic acid on brain development using artificial rearing of delta-6-desaturase knockout mice. *Prostaglandins Leukot. Essent. Fatty Acids* **2017**, *127*, 32–39. [CrossRef]
242. Contreras, M.A.; Greiner, R.S.; Chang, M.C.; Myers, C.S.; Salem, N., Jr.; Rapoport, S.I. Nutritional deprivation of alpha-linolenic acid decreases but does not abolish turnover and availability of unacylated docosahexaenoic acid and docosahexaenoyl-CoA in rat brain. *J. Neurochem.* **2000**, *75*, 2392–2400. [CrossRef] [PubMed]
243. Coronary heart disease in seven countries. Summary. *Circulation* **1970**, *41*, I186–I195.
244. Sun, G.Y.; Shelat, P.B.; Jensen, M.B.; He, Y.; Sun, A.Y.; Simonyi, A. Phospholipases A2 and inflammatory responses in the central nervous system. *Neuromolecular. Med.* **2010**, *12*, 133–148. [CrossRef]
245. Bazinet, R.P. Is the brain arachidonic acid cascade a common target of drugs used to manage bipolar disorder? *Biochem. Soc. Trans.* **2009**, *37*, 1104–1109. [CrossRef]
246. Rapoport, S.I. Arachidonic acid and the brain. *J. Nutr.* **2008**, *138*, 2515–2520. [CrossRef]
247. Duncan, R.E.; Bazinet, R.P. Brain arachidonic acid uptake and turnover: Implications for signaling and bipolar disorder. *Curr. Opin. Clin. Nutr. Metab. Care* **2010**, *13*, 130–138. [CrossRef] [PubMed]
248. Fraser, T.; Tayler, H.; Love, S. Fatty acid composition of frontal, temporal and parietal neocortex in the normal human brain and in Alzheimer's disease. *Neurochem. Res.* **2010**, *35*, 503–513. [CrossRef] [PubMed]
249. Hosono, T.; Mouri, A.; Nishitsuji, K.; Jung, C.G.; Kontani, M.; Tokuda, H.; Kawashima, H.; Shibata, H.; Suzuki, T.; Nabehsima, T.; et al. Arachidonic or docosahexaenoic acid diet prevents memory impairment in Tg2576 mice. *J. Alzheimers Dis.* **2015**, *48*, 149–162. [CrossRef]
250. Hosono, T.; Nishitsuji, K.; Nakamura, T.; Jung, C.G.; Kontani, M.; Tokuda, H.; Kawashima, H.; Kiso, Y.; Suzuki, T.; Michikawa, M. Arachidonic acid diet attenuates brain Abeta deposition in Tg2576 mice. *Brain Res.* **2015**, *1613*, 92–99. [CrossRef]
251. Katsuki, H.; Okuda, S. Arachidonic acid as a neurotoxic and neurotrophic substance. *Prog. Neurobiol.* **1995**, *46*, 607–636. [CrossRef]
252. Sanchez-Mejia, R.O.; Newman, J.W.; Toh, S.; Yu, G.Q.; Zhou, Y.; Halabisky, B.; Cisse, M.; Scearce-Levie, K.; Cheng, I.H.; Gan, L.; et al. Phospholipase A2 reduction ameliorates cognitive deficits in a mouse model of Alzheimer's disease. *Nat. Neurosci.* **2008**, *11*, 1311–1318. [CrossRef] [PubMed]
253. Vijayaraghavan, S.; Huang, B.; Blumenthal, E.M.; Berg, D.K. Arachidonic acid as a possible negative feedback inhibitor of nicotinic acetylcholine receptors on neurons. *J. Neurosci.* **1995**, *15*, 3679–3687. [CrossRef]
254. Williams, J.H.; Errington, M.L.; Lynch, M.A.; Bliss, T.V. Arachidonic acid induces a long-term activity-dependent enhancement of synaptic transmission in the hippocampus. *Nature* **1989**, *341*, 739–742. [CrossRef]
255. Fukaya, T.; Gondaira, T.; Kashiyae, Y.; Kotani, S.; Ishikura, Y.; Fujikawa, S.; Kiso, Y.; Sakakibara, M. Arachidonic acid preserves hippocampal neuron membrane fluidity in senescent rats. *Neurobiol. Aging.* **2007**, *28*, 1179–1186. [CrossRef]
256. Wang, Z.J.; Liang, C.L.; Li, G.M.; Yu, C.Y.; Yin, M. Neuroprotective effects of arachidonic acid against oxidative stress on rat hippocampal slices. *Chem. Biol. Interact.* **2006**, *163*, 207–217. [CrossRef]
257. Darios, F.; Davletov, B. Omega-3 and omega-6 fatty acids stimulate cell membrane expansion by acting on syntaxin 3. *Nature* **2006**, *440*, 813–817. [CrossRef] [PubMed]
258. Darios, F.; Ruiperez, V.; Lopez, I.; Villanueva, J.; Gutierrez, L.M.; Davletov, B. Alpha-synuclein sequesters arachidonic acid to modulate SNARE-mediated exocytosis. *EMBO Rep.* **2010**, *11*, 528–533. [CrossRef] [PubMed]
259. Yasojima, K.; Schwab, C.; McGeer, E.G.; McGeer, P.L. Distribution of cyclooxygenase-1 and cyclooxygenase-2 mRNAs and proteins in human brain and peripheral organs. *Brain Res.* **1999**, *830*, 226–236. [CrossRef]

260. McGahon, B.; Clements, M.P.; Lynch, M.A. The ability of aged rats to sustain long-term potentiation is restored when the age-related decrease in membrane arachidonic acid concentration is reversed. *Neuroscience* **1997**, *81*, 9–16. [CrossRef]

261. Angelova, P.R.; Muller, W.S. Arachidonic acid potently inhibits both postsynaptic-type Kv4.2 and presynaptic-type Kv1.4 IA potassium channels. *Eur. J. Neurosci.* **2009**, *29*, 1943–1950. [CrossRef]

262. Connell, E.; Darios, F.; Broersen, K.; Gatsby, N.; Peak-Chew, S.Y.; Rickman, C.; Davletov, B. Mechanism of arachidonic acid action on syntaxin-Munc18. *EMBO Rep.* **2007**, *8*, 414–419. [CrossRef] [PubMed]

263. Boneva, N.B.; Kikuchi, M.; Minabe, Y.; Yamashima, T. Neuroprotective and ameliorative actions of polyunsaturated fatty acids against neuronal diseases: Implication of fatty acid-binding proteins (FABP) and G protein-coupled receptor 40 (GPR40) in adult neurogenesis. *J. Pharmacol. Sci.* **2011**, *116*, 163–172. [CrossRef] [PubMed]

264. Pan, Y.; Scanlon, M.J.; Owada, Y.; Yamamoto, Y.; Porter, C.J.; Nicolazzo, J.A. Fatty acid-binding protein 5 facilitates the blood-brain barrier transport of docosahexaenoic acid. *Mol. Pharm.* **2015**, *12*, 4375–4385. [CrossRef]

265. Marszalek, J.R.; Kitidis, C.; Dirusso, C.C.; Lodish, H.F. Long-chain acyl-CoA synthetase 6 preferentially promotes DHA metabolism. *J. Biol. Chem.* **2005**, *280*, 10817–10826. [CrossRef] [PubMed]

266. Green, J.T.; Orr, S.K.; Bazinet, R.P. The emerging role of group VI calcium-independent phospholipase A2 in releasing docosahexaenoic acid from brain phospholipids. *J. Lipid Res.* **2008**, *49*, 939–944. [CrossRef]

267. Martinez, M. Tissue levels of polyunsaturated fatty acids during early human development. *J. Pediatr.* **1992**, *120*, S129–S138. [CrossRef]

268. Bitsanis, D.; Crawford, M.A.; Moodley, T.; Holmsen, H.; Ghebremeskel, K.; Djahanbakhch, O. Arachidonic acid predominates in the membrane phosphoglycerides of the early and term human placenta. *J. Nutr.* **2005**, *135*, 2566–2571. [CrossRef]

269. Dutta-Roy, A.K.; Sinha, A.K. Purification and properties of prostaglandin E1/prostacyclin receptor of human blood platelets. *J. Biol. Chem.* **1987**, *262*, 12685–12691. [CrossRef]

270. Luo, C.L.; Li, Q.Q.; Chen, X.P.; Zhang, X.M.; Li, L.L.; Li, B.X.; Zhao, Z.Q.; Tao, L.Y. Lipoxin A4 attenuates brain damage and downregulates the production of pro-inflammatory cytokines and phosphorylated mitogen-activated protein kinases in a mouse model of traumatic brain injury. *Brain Res.* **2013**, *1502*, 1–10. [CrossRef]

271. Pamplona, F.A.; Ferreira, J.; Menezes de Lima, O., Jr.; Duarte, F.S.; Bento, A.F.; Forner, S.; Villarinho, J.G.; Bellocchio, L.; Wotjak, C.T.; Lerner, R.; et al. Anti-inflammatory lipoxin A4 is an endogenous allosteric enhancer of CB1 cannabinoid receptor. *Proc. Natl. Acad. Sci. USA* **2012**, *109*, 21134–21139. [CrossRef]

272. Qawasmi, A.; Landeros-Weisenberger, A.; Leckman, J.F.; Bloch, M.H. Meta-analysis of long-chain polyunsaturated fatty acid supplementation of formula and infant cognition. *Pediatrics* **2012**. [CrossRef] [PubMed]

273. Simmer, K.; Patole, S.K.; Rao, S.C. Long-chain polyunsaturated fatty acid supplementation in infants born at term. *Cochrane Database Syst. Rev.* **2011**. [CrossRef]

274. Smithers, L.G.; Gibson, R.A.; McPhee, A.; Makrides, M. Effect of long-chain polyunsaturated fatty acid supplementation of preterm infants on disease risk and neurodevelopment: A systematic review of randomized controlled trials. *Am. J. Clin. Nutr.* **2008**, *87*, 912–920. [CrossRef]

275. Lauritzen, L.; Fewtrell, M.; Agostoni, C. Dietary arachidonic acid in perinatal nutrition: A commentary. *Pediatr. Res.* **2015**, *77*, 263–269. [CrossRef]

276. Alshweki, A.; Muñuzuri, A.P.; Baña, A.M.; de Castro, M.J.; Andrade, F.; Aldamiz-Echevarría, L.; de Pipaón, M.S.; Fraga, J.M.; Couce, M.L. Effects of different arachidonic acid supplementation on psychomotor development in very preterm infants; a randomized controlled trial. *Nutr. J.* **2015**, *14*, 101. [CrossRef] [PubMed]

277. Crawford, M.A.; Costeloe, K.; Ghebremeskel, K.; Phylactos, A.; Skirvin, L.; Stacey, F. Are deficits of arachidonic and docosahexaenoic acids responsible for the neural and vascular complications of preterm babies? *Am. J. Clin. Nutr.* **1997**, *66*, 1032S–S1041S. [CrossRef] [PubMed]

278. Forsyth, S.; Calder, P.C.; Zotor, F.; Amuna, P.; Meyer, B.; Holub, B. Dietary Docosahexaenoic Acid and Arachidonic Acid in Early Life: What Is the Best Evidence for Policymakers? *Ann. Nutr. Metab.* **2018**, *72*, 210–222. [CrossRef] [PubMed]

Review

Towards an Optimized Fetal DHA Accretion: Differences on Maternal DHA Supplementation Using Phospholipids vs. Triglycerides during Pregnancy in Different Models

Antonio Gázquez and Elvira Larqué

1   Department of Physiology, University of Murcia, 30100 Murcia, Spain; antonio.gazquez@um.es
2   Biomedical Research Institute of Murcia (IMIB-Arrixaca), 30120 Murcia, Spain
*   Correspondence: elvirada@um.es

Abstract: Docosahexaenoic acid (DHA) supplementation during pregnancy has been recommended by several health organizations due to its role in neural, visual, and cognitive development. There are several fat sources available on the market for the manufacture of these dietary supplements with DHA. These fat sources differ in the lipid structure in which DHA is esterified, mainly phospholipids (PL) and triglycerides (TG) molecules. The supplementation of DHA in the form of PL or TG during pregnancy can lead to controversial results depending on the animal model, physiological status and the fat sources utilized. The intestinal digestion, placental uptake, and fetal accretion of DHA may vary depending on the lipid source of DHA ingested by the mother. The form of DHA used in maternal supplementation that would provide an optimal DHA accretion for fetal brain development, based on the available data obtained most of them from different animal models, indicates no consistent differences in fetal accretion when DHA is provided as TG or PL. Other related lipid species are under evaluation, e.g., lyso-phospholipids, with promising results to improve DHA bioavailability although more studies are needed. In this review, the evidence on DHA bioavailability and accumulation in both maternal and fetal tissues after the administration of DHA supplementation during pregnancy in the form of PL or TG in different models is summarized.

Keywords: docosahexaenoic acid; pregnancy; supplementation; egg yolk; microalgae; placenta

Citation: Gázquez, A.; Larqué, E. Towards an Optimized Fetal DHA Accretion: Differences on Maternal DHA Supplementation Using Phospholipids vs. Triglycerides during Pregnancy in Different Models. *Nutrients* **2021**, *13*, 511. https://doi.org/10.3390/nu13020511

Academic Editors: Abdullah Mamun and Asim K. Duttaroy
Received: 18 December 2020
Accepted: 1 February 2021
Published: 4 February 2021

Publisher's Note: MDPI stays neutral with regard to jurisdictional claims in published maps and institutional affiliations.

1. Introduction

There is a growing interest in the effects of maternal diet consumed during pregnancy on both development and fetal programming of many physiological functions. During pregnancy and lactation there is an elevated docosahexaenoic acid (22:6 omega-3, DHA) requirement in the fetus and neonate as it is a critical building block of brain and retina [1–3]. In the last trimester of pregnancy, it is estimated a fetal accretion of 67 mg of omega-3 fatty acids (FA) per day, mainly DHA, and around 5% is delivered to the brain (3.1 mg/d) [4,5]. DHA conversion efficiency from α-linolenic acid (18:3 omega-3), its essential FA precursor, is very low (<1%) in fetus, placenta and newborns [6–9], being therefore insufficient to satisfy the high supply of DHA needed by the growing fetus [10,11]. Moreover, several studies have shown that supplementation with α-linolenic acid in human adults is not a good strategy to increase DHA levels, being necessary the direct supplementation with the preformed DHA molecule to observe: enhanced DHA status in blood and tissues [12], higher transfer of DHA to the fetus [13] or even to increase DHA secretion in human milk [14].

2. DHA Recommendations and Health Outcomes

2.1. DHA Intake during the Perinatal Period

Omega-3 FA intake had fallen during the 20th century; the development of the modern vegetable oil industry, the use of cereal grains and the change in eating habits have produced a remarkable disparity in the ratio of consumption of omega-6 and omega-3 FA [15,16]. Omega-6 FA consumption, mainly in the form of linoleic acid (18:2 n-6), has increased at the expense of omega-3 FA (DHA and eicosapentaenoic acid (EPA, 20:5 n-3)) in the general population [15–17]. Nowadays, the intake of DHA in developed countries with free access to food of animal origin rich in micronutrients and omega-3 FA is highly variable. There are several studies evaluating DHA consumption that warn about inadequate dietary DHA intake for many women during pregnancy [18–20] (Figure 1). It is especially alarming the situation in western countries, for example in Canada and U.S. where DHA intakes are especially low, revealing that a majority of childbearing-age and pregnant women consume less than the recommended DHA dose [21,22] (Figure 1).

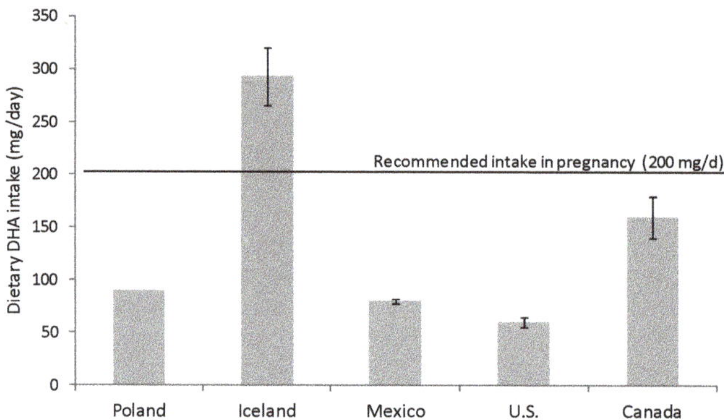

Figure 1. Estimated dietary docosahexaenoic acid (DHA) intake in pregnant women from different countries. Data from Innis and Elias 2003 [21], Parra-Cabrera et al. 2011 [18], Gunnarsdóttir et al. 2016 [23], Wierzejska et al. 2018 [24] and Zhang et al. 2018 [22].

The preferential placental uptake and transfer of DHA to the fetus in relation to other FA (palmitic, oleic and linoleic acid) has been demonstrated by the administration of stable isotope-labelled FAs to pregnant women [25,26]. Moreover, the percentage of DHA and arachidonic acid (AA; 20:4 omega-6,) both in plasma and adipose tissue is higher in the neonate than in the mother, which reveals the important role of the placenta in the concentration of these FA in the fetal compartment [27,28]. This process is known as "biomagnification" and is defined as selective enrichment of these FA in fetal, with respect to maternal plasma [29]. AA and DHA concentration in non-esterified FA (NEFA) of the intervillous space of the placenta is 3–4 times higher than in maternal blood outside the placenta [30]. This fact implies that there is certain selectivity of placental tissue for the release of these long-chain polyunsaturated FA (LC-PUFA) from the circulating lipoproteins.

It is well known that the maternal DHA intake, and hence maternal DHA levels, during pregnancy determines the DHA status of the newborn at birth and for several weeks following delivery [31–34]. Large observational studies have shown that women with low seafood intakes during pregnancy are prone to an increased risk of poor infant cognition and behavioral outcome [35,36]. Low levels of DHA and AA in maternal plasma and cord blood has been related to lower head circumference, lower birthweight, lower placental weight [32], and less cognitive and visual maturation during childhood [37,38]. Other studies found associations between omega-3 FA intake during pregnancy and lower risks of intrauterine growth restriction, preterm birth, allergies, and asthma in children [19,39,40]. However, some randomized controlled trials and meta-analysis reported inconsistent evidences and very few differences between child born from omega-3 supplemented vs. placebo mothers on long-term vision, growth and neurodevelopment outcomes [41–45]. Further follow-up studies are needed to assess the longer-term consequences and health outcomes for both mother and child of maternal omega-3 supplementation.

2.2. Dietary Recommendation during Pregnancy and Lactation

DHA dietary supplementation has been recommended by several health organizations [46,47]. European and global guidelines recommends the intake of at least 200 mg/d DHA during these periods, which can be met with two servings of fish per week [48–50]. The highest concentration of DHA is found in seafood, especially in oily fish (tuna, salmon, herring, mackerel, etc.) [51]. Smaller fishes are highly recommendable since they contain lower levels of methyl mercury and other contaminants than large-size predators [49]. Probably, the dose should be higher to detect significant effects on some outcomes but, due to the high variability in DHA intake from other sources, these recommendations are highly conservative.

DHA supplementation should be considered only if dietary consumption (natural sources) is not sufficient to meet the recommendations or when it is problematic due to food availability, socio-cultural dietary preferences, fish aversion, ethics issues (e.g., vegans), or other factors [52].

3. Lipid Sources Utilized in DHA Supplementation

The incorporation of LC-PUFA like DHA into dietary supplement products presents some technological problems because LC-PUFA are highly oxidable molecules; the large number of double bonds in their hydrocarbon chain makes advisable the addition of antioxidants or stabilizers to avoid FA oxidation. The development of microencapsulated products also allows adequate protection against oxidation, as well as the addition of these FA to powdered products [53,54]. Obtaining new fat sources with different properties, more economically competitive, healthier, or with greater bioavailability, has been an important part of the panorama of nutritional supplements for pregnant women. There are several fat sources that can be used for the manufacturing of DHA supplements. However, the lipid structure in which DHA is packaged may vary depending on the source utilized and sometimes, also on the production procedure utilized: phospholipids (PL), lyso-phospholipids (Lyso-PL), triglycerides (TG), monoglycerides, ethyl esters, etc.

3.1. Fish Oil

Fish oils are a good source of LC-PUFA omega-3 because they naturally contain high concentrations of both EPA and DHA, reaching up to 18–30% EPA + DHA in the form of TG [55]. The problems of sustainability of the large fish farms necessary for the production of these oils, typical fish odor that persists after deodorization processes and the constant increase in vegetarians and vegans contributed to the active search of alternatives for the production of this type of compounds [55,56]. In addition, the presence of some environmental contaminants in fish (e.g., methyl mercury, dioxins, and polychlorinated biphenyls), that accumulate along the marine food chain, being particularly concentrated in large predator species like shark, swordfish, or kingfish, are of special concern for pregnant women [57]. Methyl mercury is neurotoxic for the central nervous system of the fetus or newborn and its detrimental effects on neural function have been proved until young adult age [57,58]. Despite all of that, fish oil is still an important source of DHA for the production of many supplements not only for pregnant women but also for the general population.

3.2. Microalgae Oil

One of the most widespread alternatives to fish oil today is the oil obtained from culture in biofactories of different species of microalgae [56]. In fact, microalgae are the primary producers and the responsible of including LC-PUFA omega-3 in the seawater food chain [59]. Fish does not synthesize large amounts of LC-PUFA omega-3 but consumes microalgae rich in EPA and DHA or other organism fed with these microalgae [56]. Microalgae oil concentrates reach very high concentrations of DHA, containing up to 50–60% DHA in the form of TG with low levels of EPA, and have been tested in numerous investigations without any observed side effects in both animals and humans [55,60,61]. In fact, these oils are widely used in the food industry for the production of dietary supplements and the enrichment of several products [54].

3.3. Enriched Eggs

Chicken eggs are rich in protein and fat but have very little content of LC-PUFA omega-3 [62]. However, omega-3 enriched eggs can be produced by the addition of fish meal, flaxseed oil or fish oil to hen diet [54,63]; omega-3 FA increase significantly in fortified eggs resulting in up to 180 mg DHA/egg, which represent approximately the daily DHA recommended intake [64,65]. In contrast to fish and microalgae oils, egg yolk FA are distributed not only in TG but also in PL molecules (~30% of total lipids), being DHA esterified almost exclusively in phosphatidylcholine (PC) structures [66]. It has been demonstrated that consumption of omega-3 fortified eggs enhances LC-PUFA omega-3 status, including DHA levels, in breast and formula-fed infants [67] and healthy adult subjects [68,69].

Egg yolk is one of the most important sources of dietary PL of animal origin [70]. A typical Western diet contains about 3–6 g/d PL (4–8% of total fat) and a large egg can provide up to 0.8 g PL [71,72]. PC is the predominant PL species accounting for approximately 72% of the total egg PL; other PL are present in lesser quantities: 20% phosphatidylethanolamine (PE), 3% lyso-phosphatidylcholine (Lyso-PC), 3% sphingomyelin, and 2% phosphatidylinositol [70]. Beneficial health effects of dietary PL have been described over the last years related to cholesterol absorption, blood lipid profiles and cardiovascular disease risk, reduction in inflammatory processes, improvement of immunological functions, neurological development and disorders, anti-cancer properties, etc., [73,74]. However, more research is needed in order to discern what are the mechanisms involved and which effects are due to PL structure and which to the effects of FA or FA-derived metabolites carried in dietary PL [74].

3.4. Krill Oil

Krill are shrimp-like small crustaceans that live in the Antarctic ocean. Euphasia superba is the predominant species, known as Antarctic krill, and the main source of extracted krill oil [75]. Krill oil contains a considerable amount of DHA (~15%) bound to PL structures, primarily in the form of PC [75,76]. Like fish, marine microalgae are the source of LC-PUFA for krill [59]. However, in contrast to large fish, krill have a short lifespan (1–2 years) and, because they live in clean waters, are free of heavy metals, pesticides, and dioxins [77]. In the last few years, there has been a remarkable increase in the research of krill and krill oil for its health benefits in hyperlipidemia, chronic inflammation, arthritis, and premenstrual syndrome complications [75].

3.5. Lyso-Phospholipids

Lyso-PL, especially Lyso-PC, has shown to be a preferred physiological carrier of DHA to the brain and retina in some studies, being more efficiently taken than the NEFA, PL or TG form [78–82]. A similar observation has been made for erythrocytes, where DHA Lyso-PC is the major source of DHA for these cells rather than NEFA [83]. In addition, some authors have suggested that maternal erythrocytes may be a potential reserve of LC-PUFA and a preferred vehicle of them to the placenta [84]. Thus, Lyso-PL might represent an additional source of FA for the placenta. Table 1 summarizes the most relevant in vivo studies reported on Lyso-PL DHA bioavailability.

Table 1. Studies evaluating docosahexaenoic acid (DHA) bioavailability using Lyso-Phospholipids (Lyso-PL) respect to other chemical forms.

Ref.	Age	Pregnant	DHA Sources	DHA Dose	Mode of Administration	Outcomes Measured	Major Findings
	Rat/mouse models						
[79]	Young male rats	No	^{3}H-DHA as Lyso-phosphatidylcholine (Lyso-PC) and NEFA	12 nmol	Tracer infusion	^{3}H-DHA enrichment in brain, liver, kidney and heart	↑ Incorporation of ^{3}H-DHA as Lyso-PC in the brain. Similar or ↓ incorporation in other tissues compared to NEFA
[81]	Adult male mice	No	Lyso-PC and NEFA	40 mg/kg/d	Oral intake (30 days)	Plasma, liver, adipose and different brain regions fatty acids (FA). Brain function and memory tests	Lyso-PC but not NEFA increase brain DHA content. No differences in other tissues. ↑ Improvement of brain function and memory with Lyso-PC
[85]	Adult female rats	Yes	Monoacylglycerol and Lyso-PC	8 mg/kg/d	Maternal supplementation (9 weeks)	Blood, liver and adipose tissue FA in mothers. Brain regions FA in the offspring. Learning and memory skills	↑ Incorporation of DHA in cerebellum and hippocampus of pups with Lyso-PC DHA while no differences in frontal and occipital cortex. Better learning and memory scores in Lyso-PC offspring
[86]	Young male rats	No	^{14}C-DHA as Lyso-PC and NEFA	100 nmol	Tracer infusion	^{14}C-DHA enrichment in plasma, brain, heart, eyes and liver FA	↑ ^{14}C-DHA incorporation in brain after Lyso-PC administration. No differences in other tissues
[87]	Old male rats	No	^{14}C-DHA as Lyso-PC and NEFA	10 μCi	Tracer infusion	^{14}C-DHA enrichment in plasma and different brain PL pools	↓ Net rate of DHA entry into the brain with Lyso-PC ↑ ^{14}C-DHA incorporation in brain PC but ↓ in ethanolamine PL with Lyso-PC
[88]	Adult male rats	No	Lyso-PC, PL and TG	40 mg/kg/d	Oral intake (30 days)	Plasma, liver, heart, adipose tissue and different brain regions FA	Incorporation of DHA in plasma and liver: Lyso-PC > PL > TG. ↑ Incorporation of DHA from TG in heart and adipose tissue. Incorporation of DHA in brain regions: Lyso-PC > PL while no effect of DHA TG
[89]	Adult male rats	No	Lyso-PL and TG	23.5 mmol/kg diet	Oral intake (28 days)	Serum and liver FA	No differences of DHA incorporation in serum. ↑ Incorporation of DHA from TG in liver
	Human studies						
[90]	Adult men	No	^{13}C-DHA as Lyso-PC and in the form of TG	50 mg	Single oral intake	^{13}C-DHA enrichment in plasma and red blood cells PL FA	↑ ^{13}C-DHA incorporation in plasma PL with Lyso-PC. No differences in red blood cells PL

DHA; docosahexaenoic acid; FA, fatty acids; Lyso-PC, lyso-phosphatidylcholine; NEFA, non-esterified fatty acid; PL, phospholipids; TG, triglycerides. ↑ increase, ↓ decrease.

It has been demonstrated that Lyso-PL unsaturated FA at sn-2 position easily migrates to the sn-1 position in physiological conditions due to the higher reactivity of its primary alcohol [91,92]. 1-acyl Lyso-PL can be hydrolyzed by phospholipase A1, being its metabolic fate uncertain while 2-acyl is quickly reacylated and maintained in a PL structure [91]. On the other hand, during gut digestion sn-2 FA is hydrolyzed by pancreatic phospholipase A2 and that implies a lower retention in the PL structure [93]. The position of DHA within the PL molecule can affect its tissue accretion, differential incorporation of sn-1 and sn-2 Lyso-PC DHA has been reported in different brain regions in adult mice [81]. A structured 2-acyl Lyso-PL for DHA with the sn-1 position blocked (addition of an acetyl group, AceDoPC) has been synthesized in order to prevent the migration of DHA from sn-2 to sn-1 position [94]. Considering that 2-acyl Lyso-PC is the physiological form, this new molecule of DHA Lyso-PC may represent a more stable and bioavailable source of DHA for the brain and the placenta. Till now, oral intake of AceDoPC has shown good in vivo incorporation to human red blood cells PL [90,95]; additionally, it was more rapidly accumulated in brain of 20-day-old rats compared to the administration of DHA as NEFA while both compounds showed similar accretion in other tissues like plasma, heart or liver [86]. However, AceDoPC has been no tested in pregnant animals yet.

There is only one study evaluating the bioavailability of Lyso-PL form of DHA during pregnancy and it showed positive effects of maternal supplementation with DHA as Lyso-PC (obtained from egg yolk PL) in pregnant rats [85]. The authors found higher DHA content in the cerebellum and hippocampus, as well as a better score of learning and memory of pups at two months delivered by mothers supplemented with DHA Lyso-PC compared to DHA monoacylglycerides [85]. Based on available data, Lyso-PL forms of DHA might have a higher bioavailability than other lipid structures like PL, TG, or NEFA (Table 1). However, the evidences are still limited and more studies are needed to understand the biological consequences and metabolism of DHA supplementation as Lyso-PL, especially during critical periods of development like pregnancy.

3.6. Other Sources

3.6.1. Animal Products

The search for new alternative lipid sources more efficient and cost-effective for improving the intake of DHA is an area of intense research. Enriching animal products (meat, meat products, and dairy-derived food) through diet fortification with vegetable and fish sources of omega-3 FA has shown promising results in DHA concentration of edible tissues and milk. However, the inclusion of these LC-PUFA in meat presents some technological problems related to oxidative stability, off-flavors and increased production of trans and conjugated FA in the ruminants [96–99].

3.6.2. Plants

The increase in omega-3 LC-PUFA production and accumulation in plants by genetic engineering is also of interest. Plants are primary producers of essential FA (linoleic and α-linolenic acid) but lack of the natural capacity to synthesize LC-PUFA (AA, EPA, and DHA) [59]. The goal is to produce transgenic plants capable of accumulate omega-3 LC-PUFA to levels similar to that found in fish oil through the promotion of desaturase (Δ-5 and Δ-6 desaturase) and elongase (Δ6-elongase) activities [100]. Although this alternative seems very promising and may lead to significant production of omega-3 LC-PUFA for human consumption in the future, the levels of omega-3 obtained are much lower than the other types of lipid sources, the process is still expensive and the cultivation of these plants must go through a rigorous process of regulatory approvals [54,59,100].

4. Materno-Fetal Bioavailability of Different DHA Sources

The production of dietary DHA supplements has been a field of intense study and evolution in recent years. However, the use of different fat sources of DHA mean that we are taking this FA in different lipid structures, which may affect both intestinal digestion and absorption, lipoprotein distribution, metabolic fate, placental uptake and delivery to fetus [101–103]. All these observations open the door to thinking about what is the quantitative contribution of the different lipid fractions (NEFA, TG, PL, and cholesterol esters (CE)) of maternal plasma to the placenta and whether they differ in bioavailability. The incorporation of dietary FA into maternal circulating lipoproteins (mainly in the form of PL or TG) might be modified by the lipid form consumed in the diet. Therefore, it is interesting to study whether the consumption of DHA in the form of PL vs. the classical form of TG could lead to a more favorable plasma conditions for the placenta, enhancing the production of DHA-rich Lyso-PL and an increased placental uptake and fetal delivery of DHA. This would allow the design of new nutritional supplements for pregnant women with sources of DHA with higher bioavailability for the placenta. A summary of the main data from studies evaluating DHA bioavailability after TG or PL supplementation is reported in Table 2.

Table 2. Studies evaluating docosahexaenoic acid (DHA) bioavailability using triglyceride (TG) respect to phospholipid (PL) sources.

Ref.	Age	Pregnant	DHA Sources	DHA Dose	Time of Administration	Outcomes Measured	Major Findings
	Rat models						
[104]	Adult female	No	Fish oil (TG) and krill oil (PL)	1.9–4.6%	8 weeks	FA apparent digestibility and brain fatty acids (FA)	↓ Intestinal absorption and brain DHA deposition after administration as PL
[105]	Adult female	No	Tuna/fungal oil (TG) and pig brain concentrate (PL)	0.9%	3 weeks	FA excretions and fat apparent absorption	↓ Apparent absorption of DHA from pig brain PL
[105]	Adult female	No	Egg TG and egg PL	0.9%	3 weeks	FA excretions and fat apparent absorption	↑ Apparent absorption of DHA from egg PL
[106]	Adult male	No	TG and PL oils (not specified)	~1%	3 weeks	Plasma, liver and kidney FA	↓ DHA in plasma and liver after PL oil administration
[107]	Adult female	Yes	Microalgae oil (TG) and egg yolk (PL)	2.5%	3 weeks	Maternal plasma and liver FA, total fetus and fetal brain FA, placenta FA	No DHA differences in maternal plasma, fetus or placenta. ↑ DHA in maternal liver fractions with TG source
[108]	Adult female	Yes	Microalgae oil (TG), egg yolk (PL)	8 mg/kg/d	9 weeks	Maternal plasma, red blood cells, liver, adipose tissue and milk FA	No differences in maternal plasma. ↑ DHA in red blood cells and milk FA with PL source
	Pig models						
[109]	Piglets	No	Tuna/fungal oil (TG) and egg yolk (PL)	0.3%	4 weeks	Plasma and plasma lipoprotein lipid fractions FA	↑ DHA incorporation in HDL-PL fraction with egg yolk source (PL)
[110]	Piglets	No	Tuna/fungal oil (TG) and egg yolk (PL)	0.3%	16 days	Plasma FA and dry matter digestibility	↓ Intestinal absorption and plasma concentration of DHA after administration as PL
[111]	Piglets	No	Sow milk (TG) and pig brain concentrate (PL)	0.3–0.4%	17 days	Plasma PL and liver microsomes FA	↑ DHA incorporation in plasma PL and liver with DHA-PL source
[112]	Piglets	No	Fish oil (TG) and egg yolk (PL)	0.2–0.4%	2 weeks	Plasma and red blood cells FA	↑ DHA incorporation in plasma PL and with DHA-PL source
[113]	Adult female	Yes	Microalgae oil (TG) and egg yolk (PL)	0.8%	6 weeks	Maternal plasma, lipoproteins and liver FA, fetal plasma and brain FA, placenta FA	↑ DHA content in placenta with PL source but no differences in fetal tissues
	Human studies						
[114]	Preterm infants	No	Breast milk/algae oil (TG) and egg yolk (PL)	0.24–0.64%	≥5 weeks	Fecal output and FA balance	↑ Intestinal absorption of DHA administered as PL
[115]	Full term infants	No	Microalgae oil (TG) and egg yolk (PL)	0.1%	3 months	Plasma lipid fractions FA	No differences in plasma DHA
[116]	Children 8–13 y	No	Fish oil (TG) and enriched PL (not specified)	100 mg/d	3 months	Plasma and red blood cells PL fraction FA	No differences in plasma or red blood cells DHA

DHA; docosahexaenoic acid; FA, fatty acids; PL, phospholipids; TG, triglycerides. ↑ increase, ↓ decrease.

4.1. Intestinal Digestion and Absorption

The digestion of dietary fats and the subsequent FA absorption and assembly in plasma lipoproteins depends on the chemical structure in which they have been ingested (mainly TG or PL). TG digestion takes place in the small intestine where the FA of sn-1 and sn-3 positions are hydrolyzed by pancreatic lipase releasing the corresponding NEFA and 2-monoacylglycerol (Figure 2). A small part of the 2-monoacylglycerol is fully degraded to NEFA and glycerol. On the other hand, dietary PL digestion (mainly PC) occurs through the action of the pancreatic phospholipase A2 which releases the FA located at the sn-2 position generating a NEFA and a Lyso-PL, although a small part is completely hydrolyzed to NEFA and glycerol phosphocholine (Figure 2) [117,118].

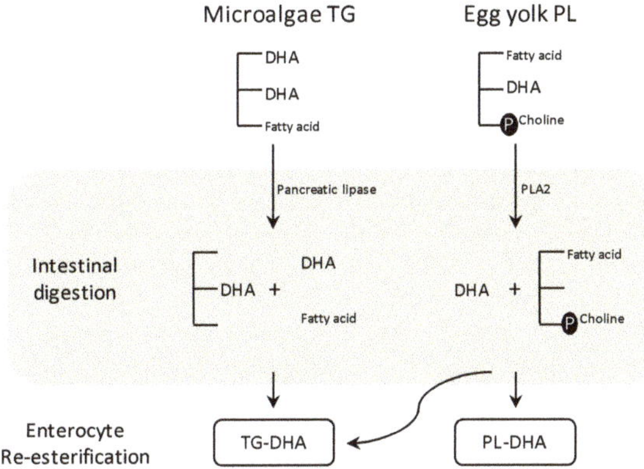

Figure 2. Intestinal digestion of dietary docosahexaenoic acid (DHA) ingested as phospholipids (PL) from egg yolk or triglycerides (TG) from microalgae oil schematic. PLA2, phospholipase A2.

The products generated during the digestion are captured by the enterocytes in a process not entirely established in which both passive diffusion processes and facilitated transport by FA binding protein associated with the plasma membrane (FABPpm), fatty acid translocase (FAT/CD36) and fatty acid transport protein (FATP) take place [101,119,120]. Once inside the enterocyte, TG and PL molecules are re-esterified and the NEFA can be incorporated into the same structure as they were part or into others that are being formed at the same time in the cell (TG, PL, or CE) (Figure 2). PL and TG captured or re-synthesized by the enterocytes are transported in the bloodstream in lipoproteins, mainly chylomicrons (QM, apolipoprotein B48) and very low density lipoproteins (VLDL, apolipoprotein B100), especially in fasting situations [121,122]. The ingestion of PL or TG modifies the diameter of secreted lipoproteins. PL ingestion produce lipoprotein with lower diameter, called by some authors "small QM", than those secreted after TG ingestion [122,123]. However, the lipid source of DHA not only affects the dimension of lipoproteins secreted but also their FA composition. The four-week administration of LC-PUFA (AA and DHA) from fungal and tuna oil resulted in a preferential incorporation of these FA in PL fraction of low density lipoprotein (LDL) particles in neonatal piglets, while when egg yolk PL were used as DHA source (DHA in the form of PL) a higher incorporation in high density lipoprotein (HDL) PL was found [109]. This has been observed not only for LC-PUFA but in general for PC which after being absorbed is preferably incorporated into HDL lipoproteins [124], probably directly in the enterocyte and without liver intervention [125].

There is controversy on the effects of PL addition to the diet on FA intestinal absorption. Several studies revealed better fat digestion and increased FA absorption when egg yolk PL were added to diets of experimental animals or infant formulas [105,107,114,126]. In fact, Carnielli et al. reported that DHA from egg PL was better absorbed in premature babies than DHA from breast milk or algae oil (TG form) [114]. However, other authors showed better absorption of FA after TG sources administration than when using PL sources (egg yolk and krill oil) in neonatal piglets and rats [104,110]. Differences in composition or the presence of other lipid components in important amounts may also affect lipid digestion and FA absorption. For example, Amate and colleagues showed higher absorption of diet supplemented with egg yolk PL compared to egg yolk TG and lower absorption of pig brain concentrate PL compared to tuna and microalgae oil [105]. The authors argued that the presence of other lipid compounds such as cerebrosides, gangliosides, esphingolipids and Lyso-PC might have affected the intestinal absorption of the experimental fats [105].

Similarly krill oil, with DHA mainly in PL, has received criticism from some experts in comparative studies of LC-PUFA omega-3 bioavailability due to its changes in composition depending on the technological processing and the season of capture [127,128]. For example, it has been described that krill oil PL content ranged from 19 to 81% and that it can contain a high amount of DHA in lipid structures different to PL such as NEFA (up to 21% DHA) and FA ethyl esters, which clearly influences both FA absorption and bioavailability [76,129–131]. Other aspects such as intestinal maturity, metabolic differences inter-species, and microflora composition influence both fat digestibility and absorption processes making difficult the comparison of FA bioavailability studies [132].

Some authors have studied the positions of the lipid structures whose FA are most likely to be preserved after the process of intestinal digestion and absorption. It has been shown that FA esterified in the sn-2 position of the TG molecule is widely conserved in TG fraction of QM particles [133,134]. On the other hand, in the case of PL, the FA esterified in the sn-1 position of PC is also protected from digestive hydrolysis [117]. Therefore, the position that occupies the DHA within the molecule is important, it could determine in which structure this DHA will later appear in the bloodstream and enhanced or minimized the expected effects of dietary supplementation.

Egg DHA shows a high stereospecificity, being almost exclusively esterified in the sn-2 position of PC and PE molecules [66,135]. This implies that a large part of the DHA from egg yolk enters the enterocyte as NEFA and not in the form of Lyso-PC and may not be maintained in a structure of PL when moving to plasma (Figure 1). In pig brain concentrate PL something similar occurs, although the composition in terms of the type of PL is not the same. Egg PL are mainly composed of PC (87% PC and 11% PE) while pig brain concentrate by PE (44% PE and 24% PC) [135]. The effects that the type of PL might have on the digestion and absorption of FA are unknown. On the other hand, the distribution of DHA within TG molecules is also not the same in all sources; tuna oil contains about 50% DHA esterified in the sn-2 position of the glycerol [135], similar values have been also described for some microalgae oils [136]. However, in other sources, such as egg yolk TG, a similar distribution has been observed between the sn-1 and sn-2 position [66]. Artificial re-esterification processes in oil contribute to a random distribution of DHA between the three available positions of glycerol [128]. Little is known about the likely influence of the characteristics of each source and the composition of the rest of FA (saturated, monounsaturated, omega-6, etc.) on the digestion and absorption of DHA. However, DHA in the sn-2 position of TG has been shown to be better absorbed than those in sn-1 or sn-3 position [137,138].

Therefore, it is more likely that DHA from microalgae oil or fish oil TG (sn-2 position TG) remains on the same TG molecule structure than DHA from egg yolk (sn-2 position PL) (Figure 2).

4.2. Circulating DHA and Metabolic Fate

The lipid fraction in which administered DHA appears in blood circulation is important since it may determine in some grade the metabolic fate and metabolism, and hence the efficiency, of the DHA supplementation applied. It is not clear whether intake of DHA during pregnancy as PL form can be a better source with higher placental bioavailability and fetal brain accretion than the consumption of DHA as TG. The use of lipid sources with PL to evaluate the bioavailability of DHA for different organs in comparison with the administration of TG sources has been extensively studied in non-pregnant humans and animals.

Studies in full-term infants, children and piglets indicated that the plasma lipid fraction in which DHA is incorporated in circulation after gut digestion and absorption is not always related with the chemical form of DHA consumed [115,116,139]. In fact, LC-PUFA are mainly incorporated in maternal plasma PL fraction while saturated and monounsaturated FA in plasma TG [25]. However, several reports in piglets showed higher circulating values of DHA in plasma PL fraction when DHA was administered as PL from different sources: Jiménez et al. in newborn formula-fed piglets after the administration of a formula enriched with LC-PUFA from pig brain concentrate compared to sow milk [111]; Amate et al. after the administration of DHA in the form of PL from egg yolk or TG from fungal and tuna oil [109]; and Alessandri et al. administered DHA-rich egg-yolk vs. fish oil [112]. On the contrary, opposite results with higher DHA enrichment in plasma after DHA-TG ingestion have been reported in rats [106] and no differences in DHA distribution between plasma lipid fractions after the supplementation with PL or TG sources the diet of full-term infants [115]. Vaisman et al. reported higher omega-3 LC-PUFA enrichment in plasma PL compared with placebo, as well as enhanced visual sustained attention score in children, but no differences were observed between PL and TG sources supplementation effects [116].

Studies in pregnant state are scarce. We supplemented the diet of pregnant rats with 2.5% DHA of total FA in the form of PL from egg yolk (mainly PC) or 2.5% DHA in the form of TG from microalgae oil and both sources produced similar values of total DHA in total serum FA and even in plasma PL fraction [107]. Valenzuela and colleagues also reported no differences in the total plasma value of DHA after the administration of this FA as PL or TG in rats at a dose of 8mg/kg/day in adult non-pregnant females, or in these same animals during gestation and after delivery [108]. Nevertheless, they observed a higher value of DHA in the erythrocyte membrane PL after delivery in animals receiving egg yolk PL compared to animals fed with microalgae oil [108]. Our group also carried out a similar experiment with pregnant sows in which animals were fed with diets containing 0.8% DHA from egg yolk (PL form) or microalgae oil (TG form) during the last third of gestation (40 d) (Figure 3) [113]. In this study, despite no differences were observed in total plasma FA profile between groups, we found a non-significant trend towards greater incorporation of DHA in plasma PL fraction in DHA-PL fed group compared to DHA-TG ($p = 0.130$), which indicates that maternal metabolism modulates in certain degree the effect of dietary DHA, modifying its incorporation in maternal serum lipid fractions [113]. The low number of animals per group may have conditioned the lack of statistical significance in DHA PL fraction data ($n = 6$/group). The higher incorporation of DHA was in HDL and LDL lipoproteins [113]. These results of higher DHA in PL fraction after the ingestion of a DHA-PL source obtained in pregnant sows were in line with previous studies performed in non-pregnant animals [109,111,112], probably some inter-species differences in metabolism and in fat sources composition influenced the discrepancies observed between rat and pig studies.

Concerning other maternal tissues (liver, adipose tissue, and brain), the supplementation with preformed DHA as PL or TG contributed equally to DHA levels either in rat or pig models [107,108,113] (Figure 3). However, higher DHA percentage was found in PL, NEFA, and CE fractions of maternal liver after DHA-TG supplementation compared to DHA supplementation from egg yolk DHA-PL in pregnant rats, revealing the high conversion of dietary DHA that takes place also in the liver [107]. Thus, not only digestion and absorption processes but also liver metabolism regulates the incorporation of DHA into different lipid fractions in maternal tissues reducing the impact of the dietary intervention with different fat sources.

We should be cautious comparing studies with different species and under different physiological conditions, since pregnancy is a special situation in which the maternal lipid metabolism is altered in order to provide all the nutrients needed for the developing fetus.

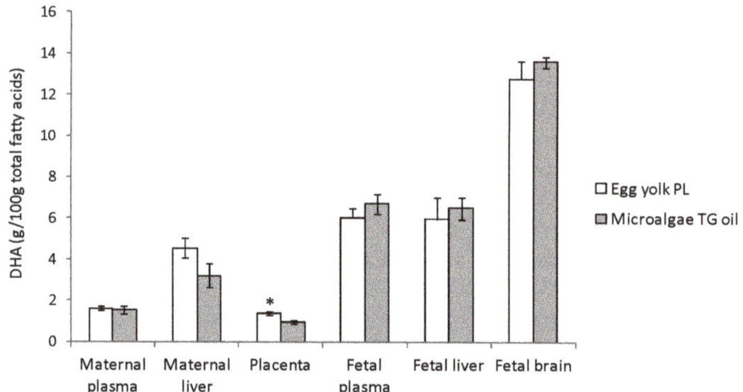

Figure 3. Docosahexaenoic acid (DHA) percentage at delivery in maternal and fetal tissues after maternal DHA supplementation (0.8% of total fatty acids) as phospholipid (PL) from egg yolk or triglycerides (TG) from microalgae oil during the last third of gestation in sows (40 d). Values are means ± SEM (n = 6/group). * Indicates significant differences between PL and TG groups of the same tissue ($p < 0.05$). Data from Gázquez et al. 2017 [113].

4.3. Placental DHA Uptake and Fetal Accretion

The process of placental FA transfer has been extensively reviewed by Larqué et al. [44,102,140]. Placenta can take FA directly from the maternal circulation in the form of NEFA, which concentration increases in the third trimester compared to a non-pregnant woman [31,102]. However, most of the FA present in maternal circulation are esterified in TG, PL, and CE structures. Placenta expresses various lipase activities responsible of the release of FA, in the form of NEFA, from circulating TG and PL, being two of them extensively studied in the literature: lipoprotein lipase (LPL) and endothelial lipase (EL) [141]. LPL preferentially hydrolyzes TG molecules carried in QM and VLDL particles releasing the corresponding NEFA for placental absorption and transfer [142,143]. On the other hand, EL breaks TG but preferably PL [144,145]. EL releases the FA in the sn-1 position of the PL yielding a NEFA and a Lyso-PL with a FA in the sn-2 position [144]. This sn-2 position of PL is usually occupied by LC-PUFA, thus EL may be able to generate DHA-rich Lyso-PL [146].

Once FA have been released, mainly in the form of NEFA, by placental lipases, they enter the cell by passive diffusion or facilitated by protein transports associated with membrane. There are different membrane proteins that have been identified so far: FABPpm, FAT/CD36 and FATP 1–6. Recently, the orphan protein named Major Facilitator Superfamily Domain Containing 2a (MFSD2a) has been described as one of the major carriers of DHA Lyso-PC [147]. When FA are inside the placenta cells, they bind to cytosolic FA binding proteins (FABP). These NEFA can also be oxidized within the trophoblast or reesterified and stored in lipid droplets. Placental cells deliver NEFA to the fetal compartment using the same FA carriers involved in FA uptake from maternal blood [148,149]. A recent study reported higher protein expression of MFSD2a carrier in the basal membrane of the syncitiothrophoblast (in contact to fetal blood) than in the microvillous membrane (in contact to maternal blood), which could indicate that placenta is also able to deliver Lyso-PL to the fetal circulation [150]. Placental metabolism plays an active role in the materno-fetal transfer of FA, combined experimental and computational modeling studies corroborated FA retention and controlled FA delivery to fetal circulation mediated by the placenta [151,152].

To our knowledge, only our research group has evaluated the incorporation of DHA to the placenta and the fetal DHA accretion after maternal supplementation with different DHA fat sources (PL and TG). The administration of 0.8% DHA in the form of PL from egg yolk to pregnant sows produced a significantly higher accumulation of DHA in placenta compared to microalgae DHA-TG (Figure 3) [113]. It may be due to higher release of DHA Lyso-PL from plasma PL lipid pool by the action of endothelial lipase in the placenta [144]. However, Lyso-DHA carrier MFSD2a did not appear to be involved since similar protein expression was observed for DHA-PL and DHA-TG fed groups [113]. This higher accumulation of DHA in placenta was exclusively in PL fraction, which makes sense since placental tissue is composed mainly by PL structures (~85%) [153]. In fact, in vivo [25] and in vitro [154] studies showed that DHA is up-taken by the placenta and esterified mostly as PL and to a lesser extent as TG.

The administration of DHA from PL or TG sources did not produce any differences in placental DHA accumulation in pregnant rats [107]. Again, the inter-species differences seem to affect the results obtained, it is important to mention that the placental structure is not the same in rats and pigs. The placenta of rodents, just like human placenta, is a discoidal endotheliochoreal placenta which represent the minimum separation (one single layer of throphoblast cells) between maternal blood and fetal capillaries while pig placenta is diffuse epitheliochoreal which higher degree of separation between both blood circulations (several layers of epithelial cells). It is unknown whether these histological differences could affect modulate FA uptake or transfer across the placenta.

Despite the discrepancies observed in the placental DHA uptake between different animal models, in all cases the accumulation of DHA in fetal organs (plasma, liver, and brain) was similar regardless of the fat source utilized for maternal DHA supplementation (egg yolk PL or microalgae TG) (Figure 3) [107,113]. Thus, the use of different lipid sources of preformed DHA in the form of PL or TG contributes to a similar fetal DHA accretion, including to the fetal brain. This means that both sources are equally efficient for the fetus.

Nevertheless, Valenzuela et al. showed higher milk DHA content of DHA-PL supplemented rats compared to DHA-TG during the first days of lactation (days 3–20 after delivery), which could have beneficial effects on fetal neurodevelopment [108]. To date no more studies evaluating milk DHA secretion after the use of different DHA sources have been published. It would be interesting to know whether milk DHA content can be increase differentially depending on the lipid source utilized since accelerated fetal brain DHA accretion in human continues up two years of life [2].

A comprehensive understanding of the actual consequences of maternal DHA supplementation during pregnancy for the fetus, especially in the case of fetal brain, is necessary to better design these DHA supplements for pregnant women. Dietary supplements for women need to be tested for real bioavailability and function in the fetus. With the evidence available today, maternal DHA supplementation as PL produces similar materno-fetal transfer of DHA across the placenta and DHA bioavailability for the fetus, including the fetal brain, than DHA in the form of TG. Intestinal digestion and absorption processes, as well as liver FA re-esterification and placental transfer seem to reduce the expected efficiency of different fat sources with DHA. Thus, both sources (PL and TG) are equally available for the developing fetus during pregnancy and can be used for the manufacture of nutritional supplements with DHA for pregnant women.

5. DHA Supplementation in Complicated Pregnancies

Some maternal diseases and conditions negatively affect LC-PUFA metabolism and their transfer across the placenta. Tomedi et al. showed that obese pregnant women were 3 times as likely of being in the lowest tertile of essential FA (DHA, EPA, and AA) [155]. Consistent with this, positive associations between maternal body mass index (BMI) and mid-pregnancy AA and omega-6 PUFA while negative association with total omega-3 PUFA has been observed in the Dutch Generation R cohort [156]. Moreover, several studies have shown reduced plasma values of DHA and AA in fetuses of mothers affected by type I diabetes [157], type II [158] and gestational diabetes mellitus (GDM) [159,160].

Our group demonstrated impaired materno-fetal transfer of DHA in GDM pregnancies by the administration of stable isotopes labeled-FA to pregnant women while the transfer of the rest of studied FA was increased [160]. Placenta of these GDM subjects showed lower expression of MFSD2a transporter compared to healthy women and this is associated with a lower DHA concentration in cord blood, which supports the contribution of Lyso-PL to materno-fetal DHA transport [161]. These results were corroborated by Soygur et al. [162], which reinforces the role that MFSD2a protein may play a role in the selective transfer of DHA not only in the brain but also in the placenta. GDM also produces changes in the expression of other FA transport proteins (e.g., FAT, adipocyte FA binding protein (A-FABP) and FATP-1) and activates the insulin signaling cascade which may promote greater transport of fat to fetus [163].

An increase in mRNA and expression levels of lipoprotein receptors in placenta (LDL and VLDL receptors) has been reported in women with GDM depending on their BMI [164]. Obese placentas has shown decreased expression of FATP-4 and increased expression of FAT which could be affecting the FA uptake and transfer process [165]. Besides, an inverse association was found between DHA level in cord blood and pre-pregnancy BMI of mothers [166]. Interestingly, we found in obese pregnant women that big placentas had lower levels of PC carrying DHA and AA in intracellular lipid depots or lipid droplets, indicating not only structural alterations but also changes in FA metabolism [167]. Nevertheless, more studies and with more subjects are needed to establish the mechanisms underlying these alterations observed in GDM or obese placentas.

Some recent studies suggest lower efficiency of dietary DHA supplementation in obese and GDM pregnancies [168,169]. Monthe-Dreze et al. showed that, despite all BMI groups of pregnant women had higher omega-3 concentrations after supplementation, obese women had attenuated changes compared to lean women, resulting in a 50% difference in the effect size [169]. Min et al. supplemented GDM women with 600mg/d DHA or high oleic acid sunflower seed oil and they observed enhanced maternal but not fetal DHA status, which may indicate that placenta lowered the effect of DHA supplementation [168].

To our knowledge, no studies evaluating the effect of different DHA sources or structures (e.g., PL vs. TG) have been conducted in complicated pregnancies. All this together with the high prevalence of diabetes and obesity increases the need for new sources with greater bioavailability of DHA for the placenta and, more importantly, for the fetus.

6. Conclusions

The animal model used to evaluate the materno-fetal transfer of DHA as PL or TG is a key issue since placental structure differs among the species used for such studies. Administration of DHA-rich PL produces a modest enrichment of DHA in PL plasma lipid fraction in piglets and pregnant sows compared to DHA-TG administration while similar or opposite results have been observed in other species. Intestinal digestion, re-esterification in both gut enterocytes and liver, as well as placental transfer processes reduce the impact of the dietary intervention with different lipid sources on fetal DHA levels. Dietary lipid utilization and bioavailability comprises several metabolic processes that are not completely understood and further research is needed. There are a limited number of studies evaluating placental and fetal accretion of DHA after the administration of different fat sources to pregnant animals (PL and TG). Despite some differences observed in placental DHA content between animal species, fetal DHA accretion and, especially, fetal brain DHA accumulation after PL or TG administration was similar. However, it is important to note that the use of animal models (rodents and pigs) in most studies might have some limitations in extrapolating results to humans. Lyso-PL have been proposed as a preferred physiological carrier of DHA to the brain, the available data on DHA Lyso-PL bioavailability with respect to other sources are promising and seem to indicate an increased DHA incorporation in some tissues but more studies are needed to evaluate their effects during pregnancy, fetal bioavailability, and long-term effects on neurodevelopment. In summary, although most of the results available were obtained in animal models, both PL and TG sources can be used for the manufacture of DHA supplements during pregnancy since they show a comparable bioavailability and promote similar DHA accretion in the fetus. The dose of DHA administered is perhaps more decisive than the fat source to increase fetal DHA status.

Author Contributions: All authors equally contributed to the conceptualization, investigation, and writing of this work. All authors have read and agreed to the published version of the manuscript.

Funding: This study was financially supported by the Spanish Government (GD-BRAIN, SAF2015-69265-C2-1-R), the Maternal and Child Health and Development Research Network (RED SAMID III, RD 16/0022/0009) and the Research Excellence Group CHRONOHEALTH (Séneca Foundation, 19899/GERM/15, Murcia).

Institutional Review Board Statement: Not applicable.

Informed Consent Statement: Not applicable.

Data Availability Statement: No new data were created or analyzed in this study. Data sharing is not applicable to this article.

Acknowledgments: The authors thank Hernández-Albaladejo I. for her support with animal maintenance and samples collection.

Conflicts of Interest: The authors declare no conflict of interest.

References

1. Clandinin, M.T.; Chappell, J.E.; Leong, S.; Heim, T.; Swyer, P.R.; Chance, G.W. Intrauterine fatty acid accretion rates in human brain: Implications for fatty acid requirements. *Early Hum. Dev.* **1980**, *4*, 121–129. [CrossRef]
2. Martinez, M. Tissue levels of polyunsaturated fatty acids during early human development. *J. Pediatr.* **1992**, *120*, S129–S138. [CrossRef]
3. Coletta, J.M.; Bell, S.J.; Roman, A.S. Omega-3 Fatty acids and pregnancy. *Rev. Obstet. Gynecol.* **2010**, *3*, 163–171. [PubMed]
4. Clandinin, M.T.; Chappell, J.E.; Heim, T.; Swyer, P.R.; Chance, G.W. Fatty acid utilization in perinatal de novo synthesis of tissues. *Early Hum. Dev.* **1981**, *5*, 355–366. [CrossRef]
5. Georgieff, M.K.; Innis, S.M. Controversial nutrients that potentially affect preterm neurodevelopment: Essential fatty acids and iron. *Pediatr. Res.* **2005**, *57*, 99R–103R. [CrossRef] [PubMed]
6. Innis, S.M. Perinatal biochemistry and physiology of long-chain polyunsaturated fatty acids. *J. Pediatr.* **2003**, *143*, S1–S8. [CrossRef]
7. Pawlosky, R.J.; Hibbeln, J.R.; Novotny, J.A.; Salem, N., Jr. Physiological compartmental analysis of alpha-linolenic acid metabolism in adult humans. *J. Lipid Res.* **2001**, *42*, 1257–1265. [CrossRef]
8. Goyens, P.L.; Spilker, M.E.; Zock, P.L.; Katan, M.B.; Mensink, R.P. Conversion of alpha-linolenic acid in humans is influenced by the absolute amounts of alpha-linolenic acid and linoleic acid in the diet and not by their ratio. *Am. J. Clin. Nutr.* **2006**, *84*, 44–53. [CrossRef] [PubMed]
9. Chambaz, J.; Ravel, D.; Manier, M.C.; Pepin, D.; Mulliez, N.; Bereziat, G. Essential fatty acids interconversion in the human fetal liver. *Biol. Neonate* **1985**, *47*, 136–140. [CrossRef] [PubMed]
10. Uauy, R.; Hoffman, D.R.; Peirano, P.; Birch, D.G.; Birch, E.E. Essential fatty acids in visual and brain development. *Lipids* **2001**, *36*, 885–895. [CrossRef] [PubMed]
11. Salem, N., Jr.; Wegher, B.; Mena, P.; Uauy, R. Arachidonic and docosahexaenoic acids are biosynthesized from their 18-carbon precursors in human infants. *Proc. Natl. Acad. Sci. USA* **1996**, *93*, 49–54. [CrossRef]
12. Burdge, G.C.; Calder, P.C. Conversion of alpha-linolenic acid to longer-chain polyunsaturated fatty acids in human adults. *Reprod. Nutr. Dev.* **2005**, *45*, 581–597. [CrossRef] [PubMed]
13. De Groot, R.H.; Hornstra, G.; van Houwelingen, A.C.; Roumen, F. Effect of alpha-linolenic acid supplementation during pregnancy on maternal and neonatal polyunsaturated fatty acid status and pregnancy outcome. *Am. J. Clin. Nutr.* **2004**, *79*, 251–260. [CrossRef] [PubMed]
14. Innis, S.M. Polyunsaturated fatty acids in human milk: An essential role in infant development. *Adv. Exp. Med. Biol.* **2004**, *554*, 27–43. [CrossRef] [PubMed]
15. Sanders, T.A. Polyunsaturated fatty acids in the food chain in Europe. *Am. J. Clin. Nutr.* **2000**, *71*, 176S–178S. [CrossRef] [PubMed]
16. Blasbalg, T.L.; Hibbeln, J.R.; Ramsden, C.E.; Majchrzak, S.F.; Rawlings, R.R. Changes in consumption of omega-3 and omega-6 fatty acids in the United States during the 20th century. *Am. J. Clin. Nutr.* **2011**, *93*, 950–962. [CrossRef] [PubMed]
17. Leaf, A.; Weber, P.C. Cardiovascular effects of n-3 fatty acids. *N. Engl. J. Med.* **1988**, *318*, 549–557. [CrossRef] [PubMed]
18. Parra-Cabrera, S.; Stein, A.D.; Wang, M.; Martorell, R.; Rivera, J.; Ramakrishnan, U. Dietary intakes of polyunsaturated fatty acids among pregnant Mexican women. *Matern. Child. Nutr.* **2011**, *7*, 140–147. [CrossRef]
19. Nordgren, T.M.; Lyden, E.; Anderson-Berry, A.; Hanson, C. Omega-3 Fatty Acid Intake of Pregnant Women and Women of Childbearing Age in the United States: Potential for Deficiency? *Nutrients* **2017**, *9*, 197. [CrossRef]
20. Loosemore, E.D.; Judge, M.P.; Lammi-Keefe, C.J. Dietary intake of essential and long-chain polyunsaturated fatty acids in pregnancy. *Lipids* **2004**, *39*, 421–424. [CrossRef] [PubMed]
21. Innis, S.M.; Elias, S.L. Intakes of essential n-6 and n-3 polyunsaturated fatty acids among pregnant Canadian women. *Am. J. Clin. Nutr.* **2003**, *77*, 473–478. [CrossRef] [PubMed]

22. Zhang, Z.; Fulgoni, V.L.; Kris-Etherton, P.M.; Mitmesser, S.H. Dietary Intakes of EPA and DHA Omega-3 Fatty Acids among US Childbearing-Age and Pregnant Women: An Analysis of NHANES 2001–2014. *Nutrients* **2018**, *10*, 416. [CrossRef] [PubMed]

23. Gunnarsdottir, I.; Tryggvadottir, E.A.; Birgisdottir, B.E.; Halldorsson, T.I.; Medek, H.; Geirsson, R.T. Diet and nutrient intake of pregnant women in the capital area in Iceland. *Laeknabladid* **2016**, *102*, 378–384. [CrossRef] [PubMed]

24. Wierzejska, R.; Jarosz, M.; Wojda, B.; Siuba-Strzelinska, M. Dietary intake of DHA during pregnancy: A significant gap between the actual intake and current nutritional recommendations. *Rocz. Panstw. Zakl. Hig.* **2018**, *69*, 381–386. [CrossRef]

25. Gil-Sanchez, A.; Larque, E.; Demmelmair, H.; Acien, M.I.; Faber, F.L.; Parrilla, J.J.; Koletzko, B. Maternal-fetal in vivo transfer of [13C]docosahexaenoic and other fatty acids across the human placenta 12 h after maternal oral intake. *Am. J. Clin. Nutr.* **2010**, *92*, 115–122. [CrossRef] [PubMed]

26. Larque, E.; Demmelmair, H.; Berger, B.; Hasbargen, U.; Koletzko, B. In vivo investigation of the placental transfer of (13)C-labeled fatty acids in humans. *J. Lipid Res.* **2003**, *44*, 49–55. [CrossRef] [PubMed]

27. Crawford, M.A.; Costeloe, K.; Ghebremeskel, K.; Phylactos, A.; Skirvin, L.; Stacey, F. Are deficits of arachidonic and docosahexaenoic acids responsible for the neural and vascular complications of preterm babies? *Am. J. Clin. Nutr.* **1997**, *66*, 1032S–1041S. [CrossRef] [PubMed]

28. Bitsanis, D.; Crawford, M.A.; Moodley, T.; Holmsen, H.; Ghebremeskel, K.; Djahanbakhch, O. Arachidonic acid predominates in the membrane phosphoglycerides of the early and term human placenta. *J. Nutr.* **2005**, *135*, 2566–2571. [CrossRef]

29. Crawford, M.A.; Hassam, A.G.; Williams, G. Essential fatty acids and fetal brain growth. *Lancet* **1976**, *1*, 452–453. [CrossRef]

30. Benassayag, C.; Mignot, T.M.; Haourigui, M.; Civel, C.; Hassid, J.; Carbonne, B.; Nunez, E.A.; Ferre, F. High polyunsaturated fatty acid, thromboxane A2, and alpha-fetoprotein concentrations at the human feto-maternal interface. *J. Lipid Res.* **1997**, *38*, 276–286. [CrossRef]

31. Lauritzen, L.; Carlson, S.E. Maternal fatty acid status during pregnancy and lactation and relation to newborn and infant status. *Matern. Child. Nutr.* **2011**, *7*, 41–58. [CrossRef] [PubMed]

32. Guesnet, P.; Pugo-Gunsam, P.; Maurage, C.; Pinault, M.; Giraudeau, B.; Alessandri, J.M.; Durand, G.; Antoine, J.M.; Couet, C. Blood lipid concentrations of docosahexaenoic and arachidonic acids at birth determine their relative postnatal changes in term infants fed breast milk or formula. *Am. J. Clin. Nutr.* **1999**, *70*, 292–298. [CrossRef] [PubMed]

33. Elias, S.L.; Innis, S.M. Infant plasma trans, n-6, and n-3 fatty acids and conjugated linoleic acids are related to maternal plasma fatty acids, length of gestation, and birth weight and length. *Am. J. Clin. Nutr.* **2001**, *73*, 807–814. [CrossRef]

34. Helland, I.B.; Saugstad, O.D.; Smith, L.; Saarem, K.; Solvoll, K.; Ganes, T.; Drevon, C.A. Similar effects on infants of n-3 and n-6 fatty acids supplementation to pregnant and lactating women. *Pediatrics* **2001**, *108*, E82. [CrossRef]

35. Hibbeln, J.R.; Davis, J.M.; Steer, C.; Emmett, P.; Rogers, I.; Williams, C.; Golding, J. Maternal seafood consumption in pregnancy and neurodevelopmental outcomes in childhood (ALSPAC study): An observational cohort study. *Lancet* **2007**, *369*, 578–585. [CrossRef]

36. Oken, E.; Wright, R.O.; Kleinman, K.P.; Bellinger, D.; Amarasiriwardena, C.J.; Hu, H.; Rich-Edwards, J.W.; Gillman, M.W. Maternal fish consumption, hair mercury, and infant cognition in a U.S. Cohort. *Environ. Health Perspect.* **2005**, *113*, 1376–1380. [CrossRef] [PubMed]

37. Innis, S.M. Fatty acids and early human development. *Early Hum. Dev.* **2007**, *83*, 761–766. [CrossRef] [PubMed]

38. Lauritzen, L.; Brambilla, P.; Mazzocchi, A.; Harslof, L.B.; Ciappolino, V.; Agostoni, C. DHA Effects in Brain Development and Function. *Nutrients* **2016**, *8*, 6. [CrossRef]

39. Emmett, P.M.; Jones, L.R.; Golding, J. Pregnancy diet and associated outcomes in the Avon Longitudinal Study of Parents and Children. *Nutr. Rev.* **2015**, *73*, 154–174. [CrossRef]

40. De Giuseppe, R.; Roggi, C.; Cena, H. n-3 LC-PUFA supplementation: Effects on infant and maternal outcomes. *Eur. J. Nutr.* **2014**, *53*, 1147–1154. [CrossRef] [PubMed]

41. Smithers, L.G.; Gibson, R.A.; Makrides, M. Maternal supplementation with docosahexaenoic acid during pregnancy does not affect early visual development in the infant: A randomized controlled trial. *Am. J. Clin. Nutr.* **2011**, *93*, 1293–1299. [CrossRef] [PubMed]

42. Makrides, M.; Gibson, R.A.; McPhee, A.J.; Yelland, L.; Quinlivan, J.; Ryan, P. Effect of DHA supplementation during pregnancy on maternal depression and neurodevelopment of young children: A randomized controlled trial. *JAMA* **2010**, *304*, 1675–1683. [CrossRef]

43. Middleton, P.; Gomersall, J.C.; Gould, J.F.; Shepherd, E.; Olsen, S.F.; Makrides, M. Omega-3 fatty acid addition during pregnancy. *Cochrane Database Syst. Rev.* **2018**, *11*, CD003402. [CrossRef]

44. Larque, E.; Demmelmair, H.; Gil-Sanchez, A.; Prieto-Sanchez, M.T.; Blanco, J.E.; Pagan, A.; Faber, F.L.; Zamora, S.; Parrilla, J.J.; Koletzko, B. Placental transfer of fatty acids and fetal implications. *Am. J. Clin. Nutr.* **2011**, *94*, 1908S–1913S. [CrossRef] [PubMed]

45. Newberry, S.J.; Chung, M.; Booth, M.; Maglione, M.A.; Tang, A.M.; O'Hanlon, C.E.; Wang, D.D.; Okunogbe, A.; Huang, C.; Motala, A.; et al. Omega-3 Fatty Acids and Maternal and Child Health: An Updated Systematic Review. *Evid. Rep. Technol. Assess* **2016**, 1–826. [CrossRef]

46. Innis, S.M. The role of dietary n-6 and n-3 fatty acids in the developing brain. *Dev. Neurosci.* **2000**, *22*, 474–480. [CrossRef]

47. Lauritzen, L.; Hansen, H.S.; Jorgensen, M.H.; Michaelsen, K.F. The essentiality of long chain n-3 fatty acids in relation to development and function of the brain and retina. *Prog. Lipid Res.* **2001**, *40*, 1–94. [CrossRef]

48. World Health Organization. Fats and fatty acids in human nutrition. In Proceedings of the Joint FAO/WHO Expert Consultation, Geneva, Switzerland, 10–14 November 2008; pp. 5–300.
49. Koletzko, B.; Cetin, I.; Brenna, J.T. Dietary fat intakes for pregnant and lactating women. *Br. J. Nutr.* **2007**, *98*, 873–877. [CrossRef]
50. Panel on Dietetic Products, Nutrition and Allergies. Scientific opinion of the Panel on Dietary Reference Values for fats, including saturated fatty acids, polyunsaturated fatty acids, monounsaturated fatty acids, trans fatty acids, and cholesterol. *EFSA J.* **2010**, *8*, 1461.
51. Lee, J.H.; O'Keefe, J.H.; Lavie, C.J.; Harris, W.S. Omega-3 fatty acids: Cardiovascular benefits, sources and sustainability. *Nat. Rev. Cardiol.* **2009**, *6*, 753–758. [CrossRef] [PubMed]
52. Kris-Etherton, P.M.; Hill, A.M. N-3 fatty acids: Food or supplements? *J. Am. Diet. Assoc.* **2008**, *108*, 1125–1130. [CrossRef]
53. Valenzuela, A.; Nieto, M.S. Acido docosahexaenoico (DHA) en el desarrollo fetal y en la nutrición materno-infantil. *Rev. Méd. Chile* **2001**, *129*, 1203–1211. [CrossRef] [PubMed]
54. Ganesan, B.; Brothersen, C.; McMahon, D.J. Fortification of foods with omega-3 polyunsaturated fatty acids. *Crit. Rev. Food Sci. Nutr.* **2014**, *54*, 98–114. [CrossRef]
55. Martins, D.A.; Custodio, L.; Barreira, L.; Pereira, H.; Ben-Hamadou, R.; Varela, J.; Abu-Salah, K.M. Alternative sources of n-3 long-chain polyunsaturated fatty acids in marine microalgae. *Mar. Drugs* **2013**, *11*, 2259–2281. [CrossRef] [PubMed]
56. Adarme-Vega, T.C.; Lim, D.K.; Timmins, M.; Vernen, F.; Li, Y.; Schenk, P.M. Microalgal biofactories: A promising approach towards sustainable omega-3 fatty acid production. *Microb. Cell Fact* **2012**, *11*, 96. [CrossRef]
57. Diez, S. Human health effects of methylmercury exposure. *Rev. Environ. Contam. Toxicol.* **2009**, *198*, 111–132. [CrossRef]
58. Antonelli, M.C.; Pallares, M.E.; Ceccatelli, S.; Spulber, S. Long-term consequences of prenatal stress and neurotoxicants exposure on neurodevelopment. *Prog. Neurobiol.* **2017**, *155*, 21–35. [CrossRef] [PubMed]
59. Adarme-Vega, T.C.; Thomas-Hall, S.R.; Schenk, P.M. Towards sustainable sources for omega-3 fatty acids production. *Curr. Opin. Biotechnol.* **2014**, *26*, 14–18. [CrossRef]
60. Kyle, D.J.; Arterburn, L.M. Single cell oil sources of docosahexaenoic acid: Clinical studies. *World Rev. Nutr. Diet.* **1998**, *83*, 116–131. [PubMed]
61. Hammond, B.G.; Mayhew, D.A.; Naylor, M.W.; Ruecker, F.A.; Mast, R.W.; Sander, W.J. Safety assessment of DHA-rich microalgae from Schizochytrium sp. *Regul. Toxicol. Pharmacol.* **2001**, *33*, 192–204. [CrossRef]
62. Rehault-Godbert, S.; Guyot, N.; Nys, Y. The Golden Egg: Nutritional Value, Bioactivities, and Emerging Benefits for Human Health. *Nutrients* **2019**, *11*, 684. [CrossRef]
63. Scheideler, S.E.; Froning, G.W. The combined influence of dietary flaxseed variety, level, form, and storage conditions on egg production and composition among vitamin E-supplemented hens. *Poult. Sci.* **1996**, *75*, 1221–1226. [CrossRef]
64. Hargis, P.S.; Van Elswyk, M.E.; Hargis, B.M. Dietary modification of yolk lipid with menhaden oil. *Poult. Sci.* **1991**, *70*, 874–883. [CrossRef]
65. Leskanich, C.O.; Noble, R.C. Manipulation of the n-3 polyunsaturated fatty acid composition of avian eggs and meat. *World Poult. Sci. J.* **1997**, *53*, 155–183. [CrossRef]
66. Schreiner, M.; Moreira, R.G.; Hulan, H.W. Positional distribution of fatty acids in egg yolk lipids. *J. Food Lipids* **2006**, *13*, 36–56. [CrossRef]
67. Makrides, M.; Hawkes, J.S.; Neumann, M.A.; Gibson, R.A. Nutritional effect of including egg yolk in the weaning diet of breast-fed and formula-fed infants: A randomized controlled trial. *Am. J. Clin. Nutr.* **2002**, *75*, 1084–1092. [CrossRef]
68. Farrell, D.J. Enrichment of hen eggs with n-3 long-chain fatty acids and evaluation of enriched eggs in humans. *Am. J. Clin. Nutr.* **1998**, *68*, 538–544. [CrossRef]
69. Jiang, Z.; Sim, J.S. Consumption of n-3 polyunsaturated fatty acid-enriched eggs and changes in plasma lipids of human subjects. *Nutrition* **1993**, *9*, 513–518. [PubMed]
70. Lordan, R.; Tsoupras, A.; Zabetakis, I. Phospholipids of Animal and Marine Origin: Structure, Function, and Anti-Inflammatory Properties. *Molecules* **2017**, *22*, 1964. [CrossRef]
71. Akoh, C.C. *Handbook of Functional Lipids*; Taylor & Francis: Boca Raton, FL, USA, 2006; p. 525.
72. Cohn, J.S.; Kamili, A.; Wat, E.; Chung, R.W.; Tandy, S. Dietary phospholipids and intestinal cholesterol absorption. *Nutrients* **2010**, *2*, 116–127. [CrossRef] [PubMed]
73. Blesso, C.N. Egg phospholipids and cardiovascular health. *Nutrients* **2015**, *7*, 2731–2747. [CrossRef] [PubMed]
74. Kullenberg, D.; Taylor, L.A.; Schneider, M.; Massing, U. Health effects of dietary phospholipids. *Lipids Health Dis.* **2012**, *11*, 3. [CrossRef]
75. Kwantes, J.M.; Grundmann, O. A brief review of krill oil history, research, and the commercial market. *J. Diet Suppl.* **2015**, *12*, 23–35. [CrossRef]
76. Ahmmed, M.K.; Ahmmed, F.; Tian, H.; Carne, A.; Bekhit, A.E.-D. Marine omega-3 (n-3) phospholipids: A comprehensive review of their properties, sources, bioavailability, and relation to brain health. *Compr. Rev. Food Sci. Food Saf.* **2020**, *19*, 64–123. [CrossRef]
77. Deutsch, L. Evaluation of the effect of Neptune Krill Oil on chronic inflammation and arthritic symptoms. *J. Am. Coll. Nutr.* **2007**, *26*, 39–48. [CrossRef] [PubMed]
78. Lagarde, M.; Bernoud, N.; Brossard, N.; Lemaitre-Delaunay, D.; Thies, F.; Croset, M.; Lecerf, J. Lysophosphatidylcholine as a preferred carrier form of docosahexaenoic acid to the brain. *J. Mol. Neurosci.* **2001**, *16*, 201–204. [CrossRef]

79. Thies, F.; Pillon, C.; Moliere, P.; Lagarde, M.; Lecerf, J. Preferential incorporation of sn-2 lysoPC DHA over unesterified DHA in the young rat brain. *Am. J. Physiol.* **1994**, *267*, R1273–R1279. [CrossRef] [PubMed]

80. Chouinard-Watkins, R.; Lacombe, R.J.S.; Metherel, A.H.; Masoodi, M.; Bazinet, R.P. DHA Esterified to Phosphatidylserine or Phosphatidylcholine is More Efficient at Targeting the Brain than DHA Esterified to Triacylglycerol. *Mol. Nutr. Food Res.* **2019**, *63*, e1801224. [CrossRef]

81. Sugasini, D.; Thomas, R.; Yalagala, P.C.R.; Tai, L.M.; Subbaiah, P.V. Dietary docosahexaenoic acid (DHA) as lysophosphatidylcholine, but not as free acid, enriches brain DHA and improves memory in adult mice. *Sci. Rep.* **2017**, *7*, 11263. [CrossRef]

82. Sugasini, D.; Yalagala, P.C.R.; Subbaiah, P.V. Efficient Enrichment of Retinal DHA with Dietary Lysophosphatidylcholine-DHA: Potential Application for Retinopathies. *Nutrients* **2020**, *12*, 3114. [CrossRef]

83. Brossard, N.; Croset, M.; Normand, S.; Pousin, J.; Lecerf, J.; Laville, M.; Tayot, J.L.; Lagarde, M. Human plasma albumin transports [13C]docosahexaenoic acid in two lipid forms to blood cells. *J. Lipid Res.* **1997**, *38*, 1571–1582. [CrossRef]

84. Ghebremeskel, K.; Min, Y.; Crawford, M.A.; Nam, J.H.; Kim, A.; Koo, J.N.; Suzuki, H. Blood fatty acid composition of pregnant and nonpregnant Korean women: Red cells may act as a reservoir of arachidonic acid and docosahexaenoic acid for utilization by the developing fetus. *Lipids* **2000**, *35*, 567–574. [CrossRef] [PubMed]

85. Valenzuela, A.; Nieto, S.; Sanhueza, J.; Morgado, N.; Rojas, I.; Zañartu, P. Supplementing female rats with DHA-lysophosphatidylcholine increases docosahexaenoic acid and acetylcholine contents in the brain and improves the memory and learning capabilities of the pups. *Grasas Aceites* **2010**, *61*, 16–23. [CrossRef]

86. Hachem, M.; Geloen, A.; Van, A.L.; Foumaux, B.; Fenart, L.; Gosselet, F.; Da Silva, P.; Breton, G.; Lagarde, M.; Picq, M.; et al. Efficient Docosahexaenoic Acid Uptake by the Brain from a Structured Phospholipid. *Mol. Neurobiol.* **2016**, *53*, 3205–3215. [CrossRef] [PubMed]

87. Chouinard-Watkins, R.; Chen, C.T.; Metherel, A.H.; Lacombe, R.J.S.; Thies, F.; Masoodi, M.; Bazinet, R.P. Phospholipid class-specific brain enrichment in response to lysophosphatidylcholine docosahexaenoic acid infusion. *Biochim. Biophys. Acta Mol. Cell Biol. Lipids* **2017**, *1862*, 1092–1098. [CrossRef]

88. Sugasini, D.; Yalagala, P.C.R.; Goggin, A.; Tai, L.M.; Subbaiah, P.V. Enrichment of brain docosahexaenoic acid (DHA) is highly dependent upon the molecular carrier of dietary DHA: Lysophosphatidylcholine is more efficient than either phosphatidylcholine or triacylglycerol. *J. Nutr. Biochem.* **2019**, *74*, 108231. [CrossRef]

89. Hosomi, R.; Fukunaga, K.; Nagao, T.; Tanizaki, T.; Miyauchi, K.; Yoshida, M.; Kanda, S.; Nishiyama, T.; Takahashi, K. Effect of Dietary Partial Hydrolysate of Phospholipids, Rich in Docosahexaenoic Acid-Bound Lysophospholipids, on Lipid and Fatty Acid Composition in Rat Serum and Liver. *J. Food Sci.* **2019**, *84*, 183–191. [CrossRef]

90. Hachem, M.; Nacir, H.; Picq, M.; Belkouch, M.; Bernoud-Hubac, N.; Windust, A.; Meiller, L.; Sauvinet, V.; Feugier, N.; Lambert-Porcheron, S.; et al. Docosahexaenoic Acid (DHA) Bioavailability in Humans after Oral Intake of DHA-Containing Triacylglycerol or the Structured Phospholipid AceDoPC((R)). *Nutrients* **2020**, *12*, 251. [CrossRef]

91. Croset, M.; Brossard, N.; Polette, A.; Lagarde, M. Characterization of plasma unsaturated lysophosphatidylcholines in human and rat. *Biochem. J.* **2000**, *345*, 61–67. [CrossRef]

92. Lo Van, A.; Sakayori, N.; Hachem, M.; Belkouch, M.; Picq, M.; Lagarde, M.; Osumi, N.; Bernoud-Hubac, N. Mechanisms of DHA transport to the brain and potential therapy to neurodegenerative diseases. *Biochimie* **2016**, *130*, 163–167. [CrossRef]

93. Subbaiah, P.V.; Dammanahalli, K.J.; Yang, P.; Bi, J.; O'Donnell, J.M. Enhanced incorporation of dietary DHA into lymph phospholipids by altering its molecular carrier. *Biochim. Biophys. Acta* **2016**, *1861*, 723–729. [CrossRef] [PubMed]

94. Polette, A.; Deshayes, C.; Chantegrel, B.; Croset, M.; Armstrong, J.M.; Lagarde, M. Synthesis of acetyl,docosahexaenoyl-glycerophosphocholine and its characterization using nuclear magnetic resonance. *Lipids* **1999**, *34*, 1333–1337. [CrossRef] [PubMed]

95. Lagarde, M.; Hachem, M.; Bernoud-Hubac, N.; Picq, M.; Vericel, E.; Guichardant, M. Biological properties of a DHA-containing structured phospholipid (AceDoPC) to target the brain. *Prostaglandins Leukot Essent Fatty Acids* **2015**, *92*, 63–65. [CrossRef] [PubMed]

96. Givens, D.I.; Kliem, K.E.; Gibbs, R.A. The role of meat as a source of n-3 polyunsaturated fatty acids in the human diet. *Meat Sci.* **2006**, *74*, 209–218. [CrossRef] [PubMed]

97. Rymer, C.; Givens, D.I. n-3 fatty acid enrichment of edible tissue of poultry: A review. *Lipids* **2005**, *40*, 121–130. [CrossRef]

98. Lee, S.A.; Whenham, N.; Bedford, M.R. Review on docosahexaenoic acid in poultry and swine nutrition: Consequence of enriched animal products on performance and health characteristics. *Anim. Nutr.* **2019**, *5*, 11–21. [CrossRef]

99. Nguyen, Q.V.; Malau-Aduli, B.S.; Cavalieri, J.; Malau-Aduli, A.E.O.; Nichols, P.D. Enhancing Omega-3 Long-Chain Polyunsaturated Fatty Acid Content of Dairy-Derived Foods for Human Consumption. *Nutrients* **2019**, *11*, 743. [CrossRef]

100. Ruiz-Lopez, N.; Usher, S.; Sayanova, O.V.; Napier, J.A.; Haslam, R.P. Modifying the lipid content and composition of plant seeds: Engineering the production of LC-PUFA. *Appl. Microbiol. Biotechnol.* **2015**, *99*, 143–154. [CrossRef]

101. Iqbal, J.; Hussain, M.M. Intestinal lipid absorption. *Am. J. Physiol. Endocrinol. Metab.* **2009**, *296*, 1183–1194. [CrossRef]

102. Gil-Sanchez, A.; Demmelmair, H.; Parrilla, J.J.; Koletzko, B.; Larque, E. Mechanisms involved in the selective transfer of long chain polyunsaturated Fatty acids to the fetus. *Front. Genet.* **2011**, *2*, 57. [CrossRef]

103. Ikeda, I.; Sasaki, E.; Yasunami, H.; Nomiyama, S.; Nakayama, M.; Sugano, M.; Imaizumi, K.; Yazawa, K. Digestion and lymphatic transport of eicosapentaenoic and docosahexaenoic acids given in the form of triacylglycerol, free acid and ethyl ester in rats. *Biochim. Biophys. Acta* **1995**, *1259*, 297–304. [CrossRef]

104. Tou, J.C.; Altman, S.N.; Gigliotti, J.C.; Benedito, V.A.; Cordonier, E.L. Different sources of omega-3 polyunsaturated fatty acids affects apparent digestibility, tissue deposition, and tissue oxidative stability in growing female rats. *Lipids Health Dis.* **2011**, *10*, 179. [CrossRef]

105. Amate, L.; Gil, A.; Ramirez, M. Dietary long-chain polyunsaturated fatty acids from different sources affect fat and fatty acid excretions in rats. *J. Nutr.* **2001**, *131*, 3216–3221. [CrossRef] [PubMed]

106. Song, J.H.; Fujimoto, K.; Miyazawa, T. Polyunsaturated (n-3) fatty acids susceptible to peroxidation are increased in plasma and tissue lipids of rats fed docosahexaenoic acid-containing oils. *J. Nutr.* **2000**, *130*, 3028–3033. [CrossRef]

107. Gazquez, A.; Hernandez-Albaladejo, I.; Larque, E. Docosahexaenoic acid supplementation during pregnancy as phospholipids did not improve the incorporation of this fatty acid into rat fetal brain compared with the triglyceride form. *Nutr. Res.* **2017**, *37*, 78–86. [CrossRef] [PubMed]

108. Valenzuela, A.; Nieto, S.; Sanhueza, J.; Nunez, M.J.; Ferrer, C. Tissue accretion and milk content of docosahexaenoic acid in female rats after supplementation with different docosahexaenoic acid sources. *Ann. Nutr. Metab.* **2005**, *49*, 325–332. [CrossRef]

109. Amate, L.; Gil, A.; Ramirez, M. Feeding infant piglets formula with long-chain polyunsaturated fatty acids as triacylglycerols or phospholipids influences the distribution of these fatty acids in plasma lipoprotein fractions. *J. Nutr.* **2001**, *131*, 1250–1255. [CrossRef]

110. Mathews, S.A.; Oliver, W.T.; Phillips, O.T.; Odle, J.; Diersen-Schade, D.A.; Harrell, R.J. Comparison of triglycerides and phospholipids as supplemental sources of dietary long-chain polyunsaturated fatty acids in piglets. *J. Nutr.* **2002**, *132*, 3081–3089. [CrossRef] [PubMed]

111. Jiménez, J.; Boza, J.; Suárez, M.D.; Gil, A. The effect of a formula supplemented with n-3 and n-6 long-chain polyunsaturated fatty acids on plasma phospholipid, liver microsomal, retinal, and brain fatty acid composition in neonatal piglets. *J. Nutr. Biochem.* **1997**, *8*, 217–223. [CrossRef]

112. Alessandri, J.M.; Goustard, B.; Guesnet, P.; Durand, G. Docosahexaenoic acid concentrations in retinal phospholipids of piglets fed an infant formula enriched with long-chain polyunsaturated fatty acids: Effects of egg phospholipids and fish oils with different ratios of eicosapentaenoic acid to docosahexaenoic acid. *Am. J. Clin. Nutr.* **1998**, *67*, 377–385.

113. Gazquez, A.; Ruiz-Palacios, M.; Larque, E. DHA supplementation during pregnancy as phospholipids or TAG produces different placental uptake but similar fetal brain accretion in neonatal piglets. *Br. J. Nutr.* **2017**, *118*, 981–988. [CrossRef]

114. Carnielli, V.P.; Verlato, G.; Pederzini, F.; Luijendijk, I.; Boerlage, A.; Pedrotti, D.; Sauer, P.J. Intestinal absorption of long-chain polyunsaturated fatty acids in preterm infants fed breast milk or formula. *Am. J. Clin. Nutr.* **1998**, *67*, 97–103. [CrossRef]

115. Sala-Vila, A.; Castellote, A.I.; Campoy, C.; Rivero, M.; Rodriguez-Palmero, M.; Lopez-Sabater, M.C. The source of long-chain PUFA in formula supplements does not affect the fatty acid composition of plasma lipids in full-term infants. *J. Nutr.* **2004**, *134*, 868–873. [CrossRef]

116. Vaisman, N.; Kaysar, N.; Zaruk-Adasha, Y.; Pelled, D.; Brichon, G.; Zwingelstein, G.; Bodennec, J. Correlation between changes in blood fatty acid composition and visual sustained attention performance in children with inattention: Effect of dietary n-3 fatty acids containing phospholipids. *Am. J. Clin. Nutr.* **2008**, *87*, 1170–1180. [CrossRef]

117. Le Kim, D.; Betzing, H. Intestinal absorption of polyunsaturated phosphatidylcholine in the rat. *Hoppe Seylers Z. Physiol. Chem.* **1976**, *357*, 1321–1331. [CrossRef]

118. Subbaiah, P.V.; Ganguly, J. Studies on the phospholipases of rat intestinal mucosa. *Biochem. J.* **1970**, *118*, 233–239. [CrossRef]

119. Chen, M.; Yang, Y.; Braunstein, E.; Georgeson, K.E.; Harmon, C.M. Gut expression and regulation of FAT/CD36: Possible role in fatty acid transport in rat enterocytes. *Am. J. Physiol. Endocrinol. Metab.* **2001**, *281*, E916–E923. [CrossRef] [PubMed]

120. Niot, I.; Poirier, H.; Tran, T.T.; Besnard, P. Intestinal absorption of long-chain fatty acids: Evidence and uncertainties. *Prog. Lipid Res.* **2009**, *48*, 101–115. [CrossRef] [PubMed]

121. Phan, C.T.; Tso, P. Intestinal lipid absorption and transport. *Front. Biosci.* **2001**, *6*, D299–D319. [CrossRef]

122. Tso, P.; Drake, D.S.; Black, D.D.; Sabesin, S.M. Evidence for separate pathways of chylomicron and very low-density lipoprotein assembly and transport by rat small intestine. *Am. J. Physiol.* **1984**, *247*, G599–G610. [CrossRef] [PubMed]

123. Amate, L.; Gil, A.; Ramirez, M. Dietary long-chain PUFA in the form of TAG or phospholipids influence lymph lipoprotein size and composition in piglets. *Lipids* **2002**, *37*, 975–980. [CrossRef] [PubMed]

124. Zierenberg, O.; Grundy, S.M. Intestinal absorption of polyenephosphatidylcholine in man. *J. Lipid Res.* **1982**, *23*, 1136–1142. [CrossRef]

125. Krimbou, L.; Hajj Hassan, H.; Blain, S.; Rashid, S.; Denis, M.; Marcil, M.; Genest, J. Biogenesis and speciation of nascent apoA-I-containing particles in various cell lines. *J. Lipid Res.* **2005**, *46*, 1668–1677. [CrossRef] [PubMed]

126. Morgan, C.; Davies, L.; Corcoran, F.; Stammers, J.; Colley, J.; Spencer, S.A.; Hull, D. Fatty acid balance studies in term infants fed formula milk containing long-chain polyunsaturated fatty acids. *Acta Paediatr.* **1998**, *87*, 136–142. [CrossRef] [PubMed]

127. Salem, N., Jr.; Kuratko, C.N. A reexamination of krill oil bioavailability studies. *Lipids Health Dis.* **2014**, *13*, 137. [CrossRef]

128. Schuchardt, J.P.; Hahn, A. Bioavailability of long-chain omega-3 fatty acids. *Prostaglandins Leukot Essent Fatty Acids* **2013**, *89*, 1–8. [CrossRef]

129. Schuchardt, J.P.; Schneider, I.; Meyer, H.; Neubronner, J.; von Schacky, C.; Hahn, A. Incorporation of EPA and DHA into plasma phospholipids in response to different omega-3 fatty acid formulations—A comparative bioavailability study of fish oil vs. krill oil. *Lipids Health Dis.* **2011**, *10*, 145. [CrossRef] [PubMed]

130. Araujo, P.; Zhu, H.; Breivik, J.F.; Hjelle, J.I.; Zeng, Y. Determination and structural elucidation of triacylglycerols in krill oil by chromatographic techniques. *Lipids* **2014**, *49*, 163–172. [CrossRef] [PubMed]
131. Srigley, C.T.; Orr-Tokle, I.C. Presence of Fatty-Acid Ethyl Esters in Krill Oil Dietary Supplements. *Lipids* **2018**, *53*, 749–754. [CrossRef] [PubMed]
132. Moughan, P.J.; Birtles, M.J.; Cranwell, P.D.; Smith, W.C.; Pedraza, M. The piglet as a model animal for studying aspects of digestion and absorption in milk-fed human infants. *World Rev. Nutr. Diet* **1992**, *67*, 40–113. [CrossRef]
133. Small, D.M. The effects of glyceride structure on absorption and metabolism. *Annu. Rev. Nutr.* **1991**, *11*, 413–434. [CrossRef]
134. Pufal, D.A.; Quinlan, P.T.; Salter, A.M. Effect of dietary triacylglycerol structure on lipoprotein metabolism: A comparison of the effects of dioleoylpalmitoylglycerol in which palmitate is esterified to the 2- or 1(3)-position of the glycerol. *Biochim. Biophys. Acta* **1995**, *1258*, 41–48. [CrossRef]
135. Amate, L.; Ramirez, M.; Gil, A. Positional analysis of triglycerides and phospholipids rich in long-chain polyunsaturated fatty acids. *Lipids* **1999**, *34*, 865–871. [CrossRef]
136. Myher, J.J.; Kuksis, A.; Geher, K.; Park, P.W.; Diersen-Schade, D.A. Stereospecific analysis of triacylglycerols rich in long-chain polyunsaturated fatty acids. *Lipids* **1996**, *31*, 207–215. [CrossRef]
137. Christensen, M.S.; Hoy, C.E.; Becker, C.C.; Redgrave, T.G. Intestinal absorption and lymphatic transport of eicosapentaenoic (EPA), docosahexaenoic (DHA), and decanoic acids: Dependence on intramolecular triacylglycerol structure. *Am. J. Clin. Nutr.* **1995**, *61*, 56–61. [CrossRef] [PubMed]
138. Carnielli, V.P.; Luijendijk, I.H.; van Goudoever, J.B.; Sulkers, E.J.; Boerlage, A.A.; Degenhart, H.J.; Sauer, P.J. Feeding premature newborn infants palmitic acid in amounts and stereoisomeric position similar to that of human milk: Effects on fat and mineral balance. *Am. J. Clin. Nutr.* **1995**, *61*, 1037–1042. [CrossRef]
139. Goustard-Langelier, B.; Guesnet, P.; Durand, G.; Antoine, J.M.; Alessandri, J.M. n-3 and n-6 fatty acid enrichment by dietary fish oil and phospholipid sources in brain cortical areas and nonneural tissues of formula-fed piglets. *Lipids* **1999**, *34*, 5–16. [CrossRef] [PubMed]
140. Gil-Sanchez, A.; Koletzko, B.; Larque, E. Current understanding of placental fatty acid transport. *Curr. Opin. Clin. Nutr. Metab. Care* **2012**, *15*, 265–272. [CrossRef] [PubMed]
141. Lindegaard, M.L.; Olivecrona, G.; Christoffersen, C.; Kratky, D.; Hannibal, J.; Petersen, B.L.; Zechner, R.; Damm, P.; Nielsen, L.B. Endothelial and lipoprotein lipases in human and mouse placenta. *J. Lipid Res.* **2005**, *46*, 2339–2346. [CrossRef] [PubMed]
142. Haggarty, P. Placental regulation of fatty acid delivery and its effect on fetal growth—A review. *Placenta* **2002**, *23*, S28–S38. [CrossRef] [PubMed]
143. Dutta-Roy, A.K. Transport mechanisms for long-chain polyunsaturated fatty acids in the human placenta. *Am. J. Clin. Nutr.* **2000**, *71*, 315S–322S. [CrossRef] [PubMed]
144. Gauster, M.; Rechberger, G.; Sovic, A.; Horl, G.; Steyrer, E.; Sattler, W.; Frank, S. Endothelial lipase releases saturated and unsaturated fatty acids of high density lipoprotein phosphatidylcholine. *J. Lipid Res.* **2005**, *46*, 1517–1525. [CrossRef] [PubMed]
145. McCoy, M.G.; Sun, G.S.; Marchadier, D.; Maugeais, C.; Glick, J.M.; Rader, D.J. Characterization of the lipolytic activity of endothelial lipase. *J. Lipid Res.* **2002**, *43*, 921–929. [CrossRef]
146. Chen, S.; Subbaiah, P.V. Phospholipid and fatty acid specificity of endothelial lipase: Potential role of the enzyme in the delivery of docosahexaenoic acid (DHA) to tissues. *Biochim. Biophys. Acta* **2007**, *1771*, 1319–1328. [CrossRef]
147. Nguyen, L.N.; Ma, D.; Shui, G.; Wong, P.; Cazenave-Gassiot, A.; Zhang, X.; Wenk, M.R.; Goh, E.L.; Silver, D.L. Mfsd2a is a transporter for the essential omega-3 fatty acid docosahexaenoic acid. *Nature* **2014**, *509*, 503–506. [CrossRef]
148. Hanebutt, F.L.; Demmelmair, H.; Schiessl, B.; Larque, E.; Koletzko, B. Long-chain polyunsaturated fatty acid (LC-PUFA) transfer across the placenta. *Clin. Nutr.* **2008**, *27*, 685–693. [CrossRef] [PubMed]
149. Larque, E.; Ruiz-Palacios, M.; Koletzko, B. Placental regulation of fetal nutrient supply. *Curr. Opin. Clin. Nutr. Metab. Care* **2013**, *16*, 292–297. [CrossRef]
150. Ferchaud-Roucher, V.; Kramer, A.; Silva, E.; Pantham, P.; Weintraub, S.T.; Jansson, T.; Powell, T.L. A potential role for lysophosphatidylcholine in the delivery of long chain polyunsaturated fatty acids to the fetal circulation. *Biochim. Biophys. Acta Mol. Cell Biol. Lipids* **2019**, *1864*, 394–402. [CrossRef]
151. Perazzolo, S.; Hirschmugl, B.; Wadsack, C.; Desoye, G.; Lewis, R.M.; Sengers, B.G. The influence of placental metabolism on fatty acid transfer to the fetus. *J. Lipid Res.* **2017**, *58*, 443–454. [CrossRef] [PubMed]
152. Gazquez, A.; Prieto-Sanchez, M.T.; Blanco-Carnero, J.E.; van Harskamp, D.; Perazzolo, S.; Oosterink, J.E.; Demmelmair, H.; Schierbeek, H.; Sengers, B.G.; Lewis, R.M.; et al. In vivo kinetic study of materno-fetal fatty acid transfer in obese and normal weight pregnant women. *J. Physiol.* **2019**. [CrossRef]
153. Klingler, M.; Demmelmair, H.; Larque, E.; Koletzko, B. Analysis of FA contents in individual lipid fractions from human placental tissue. *Lipids* **2003**, *38*, 561–566. [CrossRef]
154. Johnsen, G.M.; Weedon-Fekjaer, M.S.; Tobin, K.A.; Staff, A.C.; Duttaroy, A.K. Long-chain polyunsaturated fatty acids stimulate cellular fatty acid uptake in human placental choriocarcinoma (BeWo) cells. *Placenta* **2009**, *30*, 1037–1044. [CrossRef]
155. Tomedi, L.E.; Chang, C.C.; Newby, P.K.; Evans, R.W.; Luther, J.F.; Wisner, K.L.; Bodnar, L.M. Pre-pregnancy obesity and maternal nutritional biomarker status during pregnancy: A factor analysis. *Public Health Nutr.* **2013**, *16*, 1414–1418. [CrossRef]

156. Vidakovic, A.J.; Gishti, O.; Voortman, T.; Felix, J.F.; Williams, M.A.; Hofman, A.; Demmelmair, H.; Koletzko, B.; Tiemeier, H.; Jaddoe, V.W.; et al. Maternal plasma PUFA concentrations during pregnancy and childhood adiposity: The Generation R Study. *Am. J. Clin. Nutr.* **2016**, *103*, 1017–1025. [CrossRef]

157. Ghebremeskel, K.; Thomas, B.; Lowy, C.; Min, Y.; Crawford, M.A. Type 1 diabetes compromises plasma arachidonic and docosahexaenoic acids in newborn babies. *Lipids* **2004**, *39*, 335–342. [CrossRef] [PubMed]

158. Min, Y.; Lowy, C.; Ghebremeskel, K.; Thomas, B.; Offley-Shore, B.; Crawford, M. Unfavorable effect of type 1 and type 2 diabetes on maternal and fetal essential fatty acid status: A potential marker of fetal insulin resistance. *Am. J. Clin. Nutr.* **2005**, *82*, 1162–1168. [CrossRef]

159. Thomas, B.A.; Ghebremeskel, K.; Lowy, C.; Offley-Shore, B.; Crawford, M.A. Plasma fatty acids of neonates born to mothers with and without gestational diabetes. *Prostaglandins Leukot Essent Fatty Acids* **2005**, *72*, 335–341. [CrossRef] [PubMed]

160. Pagan, A.; Prieto-Sanchez, M.T.; Blanco-Carnero, J.E.; Gil-Sanchez, A.; Parrilla, J.J.; Demmelmair, H.; Koletzko, B.; Larque, E. Materno-fetal transfer of docosahexaenoic acid is impaired by gestational diabetes mellitus. *Am. J. Physiol. Endocrinol. Metab.* **2013**, *305*, E826–E833. [CrossRef] [PubMed]

161. Prieto-Sanchez, M.T.; Ruiz-Palacios, M.; Blanco-Carnero, J.E.; Pagan, A.; Hellmuth, C.; Uhl, O.; Peissner, W.; Ruiz-Alcaraz, A.J.; Parrilla, J.J.; Koletzko, B.; et al. Placental MFSD2a transporter is related to decreased DHA in cord blood of women with treated gestational diabetes. *Clin. Nutr.* **2017**, *36*, 513–521. [CrossRef]

162. Soygur, B.; Sati, L.; Demir, R. Altered expression of human endogenous retroviruses syncytin-1, syncytin-2 and their receptors in human normal and gestational diabetic placenta. *Histol. Histopathol.* **2016**, *31*, 1037–1047. [CrossRef] [PubMed]

163. Ruiz-Palacios, M.; Prieto-Sanchez, M.T.; Ruiz-Alcaraz, A.J.; Blanco-Carnero, J.E.; Sanchez-Campillo, M.; Parrilla, J.J.; Larque, E. Insulin Treatment May Alter Fatty Acid Carriers in Placentas from Gestational Diabetes Subjects. *Int. J. Mol. Sci.* **2017**, *18*, 1203. [CrossRef] [PubMed]

164. Dube, E.; Ethier-Chiasson, M.; Lafond, J. Modulation of cholesterol transport by insulin-treated gestational diabetes mellitus in human full-term placenta. *Biol. Reprod.* **2013**, *88*, 16. [CrossRef] [PubMed]

165. Dube, E.; Gravel, A.; Martin, C.; Desparois, G.; Moussa, I.; Ethier-Chiasson, M.; Forest, J.C.; Giguere, Y.; Masse, A.; Lafond, J. Modulation of fatty acid transport and metabolism by maternal obesity in the human full-term placenta. *Biol. Reprod.* **2012**, *87*, 1–8. [CrossRef] [PubMed]

166. Cinelli, G.; Fabrizi, M.; Rava, L.; Ciofi Degli Atti, M.; Vernocchi, P.; Vallone, C.; Pietrantoni, E.; Lanciotti, R.; Signore, F.; Manco, M. Influence of Maternal Obesity and Gestational Weight Gain on Maternal and Foetal Lipid Profile. *Nutrients* **2016**, *8*, 368. [CrossRef]

167. Gazquez, A.; Uhl, O.; Ruiz-Palacios, M.; Gill, C.; Patel, N.; Koletzko, B.; Poston, L.; Larque, E. Placental lipid droplet composition: Effect of a lifestyle intervention (UPBEAT) in obese pregnant women. *Biochim. Biophys. Acta Mol. Cell Biol. Lipids* **2018**, *1863*, 998–1005. [CrossRef] [PubMed]

168. Min, Y.; Djahanbakhch, O.; Hutchinson, J.; Eram, S.; Bhullar, A.S.; Namugere, I.; Ghebremeskel, K. Efficacy of docosahexaenoic acid-enriched formula to enhance maternal and fetal blood docosahexaenoic acid levels: Randomized double-blinded placebo-controlled trial of pregnant women with gestational diabetes mellitus. *Clin. Nutr.* **2016**, *35*, 608–614. [CrossRef]

169. Monthe-Dreze, C.; Penfield-Cyr, A.; Smid, M.C.; Sen, S. Maternal Pre-Pregnancy Obesity Attenuates Response to Omega-3 Fatty Acids Supplementation During Pregnancy. *Nutrients* **2018**, *10*, 1908. [CrossRef] [PubMed]

 *nutrients*

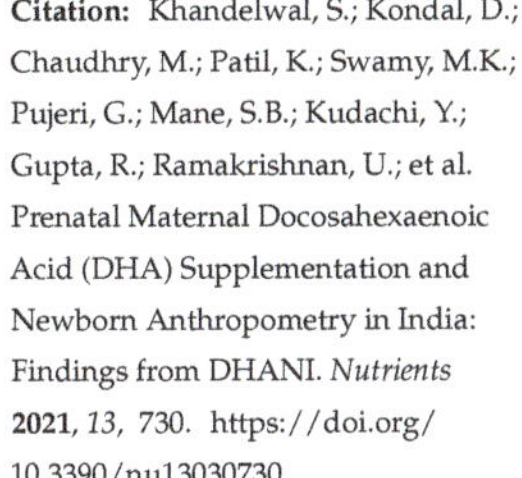

Article

Prenatal Maternal Docosahexaenoic Acid (DHA) Supplementation and Newborn Anthropometry in India: Findings from DHANI

Shweta Khandelwal, Dimple Kondal, Monica Chaudhry, Kamal Patil, Mallaiah Kenchaveeraiah Swamy, Gangubai Pujeri, Swati Babu Mane, Yashaswi Kudachi, Ruby Gupta, Usha Ramakrishnan, Aryeh D. Stein, Dorairaj Prabhakaran and Nikhil Tandon

1   Public Health Foundation of India, Gurugram 122003, India; dimple@ccdcindia.org (D.K.); monica.chaudhry@phfi.org (M.C.); ruby.gupta@phfi.org (R.G.); dprabhakaran@phfi.org (D.P.)
2   Centre for Chronic Disease Control, New Delhi 110016, India
3   Department of Obstetrics and Gynaecology, KAHER's J. N. Medical College, Belagavi 590010, India; kamalpatil1967@yahoo.co.in (K.P.); mkswamy53@yahoo.co.in (M.K.S.); pujerig@yahoo.in (G.P.); maneswati43@gmail.com (S.B.M.); yash.kudachi@gmail.com (Y.K.)
4   Hubert Department of Global Health, Rollins School of Public Health, Emory University, Atlanta, GA 30322, USA; uramakr@emory.edu (U.R.); aryeh.stein@emory.edu (A.D.S.)
5   All India Institute of Medical Sciences, New Delhi 110016, India; Nikhil_tandon@hotmail.com
*   Correspondence: shweta.khandelwal@phfi.org

Citation: Khandelwal, S.; Kondal, D.; Chaudhry, M.; Patil, K.; Swamy, M.K.; Pujeri, G.; Mane, S.B.; Kudachi, Y.; Gupta, R.; Ramakrishnan, U.; et al. Prenatal Maternal Docosahexaenoic Acid (DHA) Supplementation and Newborn Anthropometry in India: Findings from DHANI. *Nutrients* **2021**, *13*, 730. https://doi.org/10.3390/nu13030730

Academic Editor: Asim K. Duttaroy

Received: 10 January 2021
Accepted: 17 February 2021
Published: 25 February 2021

Publisher's Note: MDPI stays neutral with regard to jurisdictional claims in published maps and institutional affiliations.

Abstract: Long-chain omega-3 fatty acid status during pregnancy may influence newborn anthropometry and duration of gestation. Evidence from high-quality trials from low- and middle-income countries (LMICs) is limited. We conducted a double-blind, randomized, placebo-controlled trial among 957 pregnant women (singleton gestation, 14–20 weeks' gestation at enrollment) in India to test the effectiveness of 400 mg/day algal docosahexaenoic acid (DHA) compared to placebo provided from enrollment through delivery. Among 3379 women who were screened, 1171 were found eligible; 957 were enrolled and were randomized. The intervention was two microencapsulated algal DHA ($200 \times 2 = 400$ mg/day) or two microencapsulated soy and corn oil placebo tablets to be consumed daily from enrollment ($\leq$20 weeks) through delivery. The primary outcome was newborn anthropometry (birth weight, length, head circumference). Secondary outcomes were gestational age and 1 and 5 min Appearance, Pulse, Grimace, Activity, and Respiration (APGAR) score. The groups (DHA; $n = 478$ and placebo; $n = 479$) were well balanced at baseline. There were 902 live births. Compliance with the intervention was similar across groups (DHA: 88.5%; placebo: 87.1%). There were no significant differences between DHA and placebo groups for birth weight (2750.6 ± 421.5 vs. 2768.2 ± 436.6 g, $p = 0.54$), length (47.3 ± 2.0 vs. 47.5 ± 2.0 cm, $p = 0.13$), or head circumference (33.7 ± 1.4 vs. 33.8 ± 1.4 cm, $p = 0.15$). The mean gestational age at delivery was similar between groups (DHA: 38.8 ± 1.7 placebo: 38.8 ± 1.7 wk, $p = 0.54$) as were APGAR scores at 1 and 5 min. Supplementing mothers through pregnancy with 400 mg/day DHA did not impact the offspring's birthweight, length, or head circumference.

Keywords: docosahexaenoic acid (DHA); long chain omega-3 fatty acids; maternal supplementation; pregnancy outcomes; anthropometry; birth weight; birth length; head circumference

1. Introduction

Birth weight is a key predictor of the health trajectory of a child [1]. In 2015, the global prevalence of low birth weight (LBW) was recorded to be 14.6%, and 91% of these were from low- and middle-income countries (LMICs), primarily in southern Asia (48%) and sub-Saharan Africa (24%) [2]. LBW and preterm birth are leading causes of neonatal death in LMICs [3]. In addition, LBW is associated with an increased risk of numerous adverse health outcomes in childhood [4,5] and adulthood [6,7]. Women in deprived socio-economic conditions frequently have poor nutrition and consequently deliver infants with LBW [8]. Evidence from several studies, including birth cohorts in Brazil, Guatemala, India, The Philippines, and South Africa [9], shows that poor fetal growth carries a higher risk of chronic diseases related to nutrition later in adult life.

LBW can be the result of preterm birth (PTB) and/or intrauterine growth restriction (IUGR). The underlying causes of both PTB and IUGR are multi-factorial, including infectious diseases, hypertensive disorders, trauma and illness, maternal characteristics, and social determinants. However, the etiologies lead to a common pathway of insufficient uterine–placental perfusion and fetal nutrition [10]. Among the maternal characteristics, maternal nutritional status has been identified as one of the key determinants for LBW in India [11]. Current dietary recommendations for pregnant women emphasize protein, energy, vitamin, and mineral adequacy, but increasing attention is being given to dietary lipids, especially essential fatty acids (EFAs) [12]. Long-chain polyunsaturated fatty acid (LC-PUFA) intake during pregnancy influences both maternal and infant fatty acid status at birth [13], which itself is associated with birth weight and gestational age at birth [14]. A substantial proportion of the Indian population is vegetarian (35%, ranging from 10% to 62% across regions) or observes religious dietary restrictions that can result in multiple nutrient deficiencies [15]. Since the main dietary source of DHA is oily fish, non-supplemented vegetarian diets contain little DHA, and vegan diets contain virtually none. Indian women have low intakes of omega-3 fatty acids—median alpha linolenic acid (ALA), eicosapentanoic acid (EPA) and docosahexaenoic acid (DHA) levels are 560, 3, and 1.1 mg/day during pregnancy, respectively [16]. This is significantly lower than the daily EPA and DHA consumption recommended by the Food and Agriculture Organization (FAO) [17] (2010), for pregnant and lactating women (300 mg per day EPA + DHA, of which 200 mg per day is DHA).

Growing evidence suggests that supplementation during pregnancy with omega-3 fatty acids, especially DHA, may improve birth outcomes. In a prospective cohort study from southern India, women who did not eat fish during the third trimester had a significantly higher risk of LBW (OR: 2.49, $p = 0.019$) when compared to women whose intake was above median, that is, 9.33 g/day (interquartile range: 5.10–15.69) [18]. A review by Makrides and Best [19], documenting the global evidence on epidemiological studies and trials conducted in this area, suggested that N-3 LCPUFA supplementation during pregnancy increased the mean duration of gestation by 2 days; there was also a 40–50% reduction in early preterm birth (<34 weeks' gestation) [19]. In the United States of America, DHA supplementation resulted in longer gestation duration (2.9 d; $p = 0.041$) and greater birth weight (172 g; $p = 0.004$), length (0.7 cm; $p = 0.022$), and head circumference (HC) (0.5 cm; $p = 0.012$) [20]. Among Mexican women randomized to 400 mg/day of algal DHA or placebo from 18 to 22 weeks of gestation through delivery, the intent-to-treat analysis showed no differences between the placebo and DHA groups in newborn anthropometry, but offspring of supplemented primigravidae were 99.4 g heavier (95% CI, 5.5 to 193.4) and had 0.5 cm larger HC (diff = 95% CI, 0.1 to 0.9) than controls [21]. In the DHA to Optimize Mother Infant Outcome (DOMInO) trial from Australia, women who received fish oil supplements had a lower risk of very preterm birth (1.09% in the DHA group compared to 2.25% in the control group); mean birth weight was 68 g (95% CI, 23–114 g) heavier, and fewer infants had LBW (3.41% vs. 5.27%; 95% CI, 0.44–0.96) [22].

As results have been inconsistent, and little research on this question comes from LMIC contexts where the underlying nutritional status and etiology of LBW may differ, we assessed the impact of maternal DHA supplementation on newborn anthropometry, APGAR score, duration of gestation, and low birth weight among Indian women.

2. Materials and Methods

2.1. Trial Design and Setting

DHANI (effect of *n*-3 fatty acid (DHA) supplementation during pregnancy on newborn birth weight and gestational age in India) was established as a randomized, double-blinded, placebo-controlled trial to assess the effect of 400 mg/day algal prenatal DHA consumption by healthy Indian women from $\leq$20 weeks of singleton gestation till delivery on their offspring's size (weight, length, and head circumference) at birth. The detailed trial protocol has been published elsewhere [23]. DHANI is registered on the CTRI website as CTRI/2013/04/003540 and at clinical trials.gov as NCT01580345. Ethical clearance was obtained from institutional review boards (IRBs) of all participating institutions: Center for Chronic Disease Control (CCDC-IEC_04_2015), Public Health Foundation of India (PHFI) (TRC-IEC-261/15), and Jawaharlal Nehru Medical College (MDC/IECHSR/2016-17/A-85).

2.2. Participants and Trial Procedures

The study population was healthy pregnant women, aged 18–35 years with singleton pregnancy under $\leq$20 weeks of gestation, with no obstetric high-risk conditions, medical complications, or chronic diseases, attending the Department of Obstetrics and Gynecology at the Prabhakar Kore Hospital (PKH) in Belgavi, a largely rural district in Karnataka State, southwest India for antenatal care. Designated project staff approached women, and the consulting obstetrician on site, considering obstetric history and complications, affirmed final eligibility. Consenting eligible women were randomized by project staff to receive either 400 mg/day DHA or a placebo after providing written informed consent using a form in their preferred local language (Kannada, Marathi, or Hindi) and observed by a witness. Information on sociodemographic characteristics, obstetric and medical history, dietary intake (with a pre-piloted semi-quantitative food frequency questionnaire focusing on *n*-3 LC-PUFA-rich Indian foods), anthropometric measurements, a non-fasting blood draw, and vital signs were obtained at enrollment. The women were then given the supplements in the form of coded bottles (each bottle had a 2 week supply) matching the allotted code for the participant. Further supplements were either collected by the women from the study site or were delivered to the women's homes every fortnight by fieldworkers.

Research staff maintained contact with all women, especially during the last trimester, and visited the woman in the delivery ward within 24 h of delivery to collect data on gestational age at delivery, type of delivery, complications (if any), pregnancy outcome, APGAR (Appearance, Pulse, Grimace, Activity, Respiration) score, newborn anthropometry (weight, length, and head circumference), and maternal and cord blood samples.

2.3. Randomization, Masking, and Intervention

The randomization list for 1200 women was generated using a permuted block design (randomly allocating 600 women to DHA or placebo). The assignment code list was placed in a sealed envelope at the beginning of the study and in a secure location at PHFI by a staff member not involved in the trial. Study participants and research staff (including those at the study site) remained blinded to the treatment allocation throughout the duration of fieldwork. After obtaining due approval from the Data Safety Monitoring Board (DSMB) of the study, full analyses were carried out. Unblinding of the treatment group was done only after the generation of the primary tables.

The details of the intervention have been published already [23], but briefly, the intervention was comprised of 635 mg soft gel capsules having either 200 mg/day algal DHA or a placebo (soy/corn oil in a 50:50 ratio), identical in taste and appearance. The active ingredient DHA-S (also known as "DHA algal oil") is a naturally occurring, microalgal oil derived from *Schizochytrium* sp. (DSM Nutritional Products, Columbia, MD, USA). The sealed capsules had a shelf life of 2 years from the date of manufacture when stored at room temperature (25 °C) and 90 days once the bottle was opened. The women were instructed to store capsules in a cool, dry place and to take two capsules daily, preferably at the same time each day. Supplements were provided for more than two weeks in cases where the woman shared plans to travel. Enrolled women received supplements from the date of randomization through 6 months postpartum; for the present analysis, only supplement intake through delivery was considered.

2.4. Outcomes

The primary outcome for the DHANI trial was newborn anthropometry (birth weight, birth length, head circumference). Secondary outcomes included gestational age, APGAR scores at 1 and 5 min, still births, LBW, and preterm. All research staff at the study site were apprised of the data collection methods before the start of the trial and were provided regular refresher training every 6 months. Abstracted data included gestational age, pregnancy outcome (live birth, sex of baby, type of delivery), and APGAR score at 1 min and 5 min. Gestational age at delivery was calculated in weeks by noting the number of days from the last menstrual period (LMP) until delivery. Preterm delivery was defined as delivery after 20 weeks and before 37 completed weeks. Anthropometric data were collected by a trained research assistant within 24 h of delivery. Birth weight was measured to the nearest 10 g by using a portable single-pan digital pediatric weighing scale. Low birth weight was defined as recorded birth weight less than 2500 g. Birth length and head circumference were measured by trained research staff to the nearest 1 mm using a portable anthropometer with a fixed headpiece and a non-stretchable measuring tape, respectively, according to standard procedures. Fetal losses during pregnancy—including miscarriages/abortions and still-births and the APGAR scores were obtained from the hospital records by study personnel on-site, or details were brought by field workers (in case mother went to any other hospital). Stillbirths were defined as fetuses delivered at 20 weeks of gestation or later with no signs of life and recorded as occurring before or during the onset of labor; neonatal deaths were defined as deaths among live-born infants occurring within 28 days after delivery.

2.5. Adherence and Follow Up

Subjects were asked to maintain a daily record of their supplement consumption using a form provided by study staff. Weekly calls were made by the research staff to encourage compliance and inquire about general well-being. The used bottles were collected (for pill count) by the field-workers during the fortnightly home visits. The compliance was calculated as the total number of capsules actually consumed, expressed as a percentage of the total number expected to be consumed, which was assessed based on a compliance form filled by the participant and verified by the research staff at all home visits. A sub-sample of venous blood samples collected from the mother at recruitment and delivery was analyzed for DHA levels.

2.6. Statistical Analysis

Using data from published literature from another developing country setting [21], we estimated that a sample of 350 mothers per group would have at least 80% power to detect an effect size of 0.20 standard deviation (SD) or greater for the primary outcomes (birth weight and gestational age) at the end of the study, with a significance level of 0.05 for a two-tailed test. A 10% loss to follow-up during pregnancy and 45 as the neonatal mortality rate (NMR) were taken into account. This sample size would also allow us to detect minimum differences in birth weight of 100 g (0.2 SD) between groups with at least 80% power.

Baseline maternal and offspring characteristics were summarized as means and standard deviations or medians and inter-quartile ranges as appropriate, and categorical variables were summarized using proportions.

We used a two-sample t-test to compare the differences in mean birth weight, birth length, head circumference, and APGAR score at 1 min and 5 min at delivery between the DHA and placebo groups. We also calculated the z score for birth weight, length, and head circumference using standards established by the (International Fetal and Newborn Growth Consortium for the 21st century (INTERGROWTH-21) Project [24] and compared the difference in z score between DHA and placebo group using two-sample t-tests. The differences in proportion for preterm birth and LBW between the DHA and placebo groups were compared using the two-proportion z-test. The analysis was done using the intent to treat (ITT) principle.

We conducted several pre-specified subgroup analyses to estimate the treatment effects within different categories of maternal age (18–20, 21–25, 26–30, 31–35 years), body mass index (BMI) at enrollment (<18.5 kg/m^2; 18.5–23.0 kg/m^2; 23.0–27.5 kg/m^2; and 27.5 kg/m^2) as per Asian cut-offs [25], gravidity (multi-gravida, primi gravida), gestational age at delivery (<37, ≥37 weeks), compliance ($<80.0\%$, $\geq80.0\%$), vegetarian diet (yes, no), and child sex (male, female). The *p*-value for heterogeneity was calculated by including the interaction term between the characteristic of interest and treatment group in the linear regression model. The significance of within-subgroup treatment effects was adjusted for multiplicity for multiple subgroup analyses using the Bonferroni criterion, i.e., by dividing the overall significance level by the total number of subgroup analyses performed. For sensitivity analysis, we compared the baseline characteristics between the final study sample and those who were lost to follow-up. *p* values <0.05 were considered to be statistically significant. All statistical analysis was done using STATA 16.0 version (College Station, TX, USA) and R 3.6.2 software (Free Software Foundation, Inc., Boston, MA, USA).

3. Results

3.1. Trial Population

A total of 3379 women were screened, and 1131 were found to be eligible. Among these, 957 mothers provided informed consent and were randomly assigned to receive DHA ($n = 478$) or placebo ($n = 479$) (Figure 1).

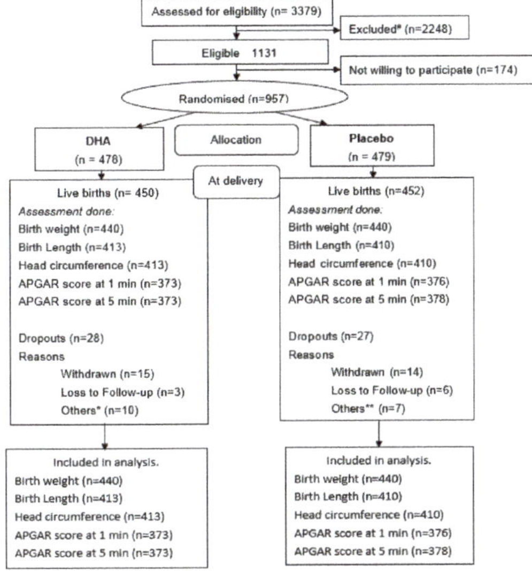

Figure 1. Consort [#] reasons for exclusion: gestational diabetes ($n = 69$); Hb < 7 g% ($n = 46$); gestational age > 20 weeks ($n = 673$); high risk pregnancies ($n = 118$); chronic conditions ($n = 246$); under any other trial ($n = 4$); delivery plan other than PK ($n = 835$); missing/wrong contact information ($n = 257$). * Others included abortion ($n = 1$); abruptio placenta ($n = 1$); fresh still birth ($n = 4$); macerated still birth ($n = 3$); neonatal death ($n = 1$) in DHA group. ** Others included fresh still birth ($n = 4$); macerated still birth ($n = 2$); medical termination ($n = 1$) in placebo group.

Overall, the mean (SD) age of the mothers was 23.5 (3.6) years, and gestational age (median (interquartile interval)) at enrolment was 15.0 (12.0, 18.0) weeks. A total of 79% of the women had completed at least secondary school and 23% of the women were employed. About 12% of the women reported monthly household income more than Rs 20,000 (285 USD taking 1 USD = 70 INR). Baseline characteristics of the enrolled women were similar between DHA and placebo groups (Table 1). The details have been published elsewhere [23]. In addition, there was no difference in baseline characteristics between those who were followed up till delivery and those who were not (Supplementary Materials Table S1).

Table 1. Maternal anthropometrics and DHA level according to treatment group at randomization.

Variable	DHA (*n* = 478)	Placebo (*n* = 479)
Maternal age (year), mean $\pm$ SD	23.5 $\pm$ 3.5	23.6 $\pm$ 3.7
Gestational age at enrollment (weeks), median (p25, p75)	15.0 (12.0, 18.0)	14.0 (12.0, 18.0)
Weight (kg), mean $\pm$ SD	48.9 $\pm$ 9.0	48.9 $\pm$ 8.5
Height (cm), mean $\pm$ SD	154.1 $\pm$ 5.6	153.9 $\pm$ 5.7
Body mass index (kg/m^2), mean $\pm$ SD	20.5 $\pm$ 3.5	20.7 $\pm$ 3.6
MUAC (cm), mean $\pm$ SD	24.3 $\pm$ 3.0	24.3 $\pm$ 3.1
Hb (g%), mean $\pm$ SD	11.1 $\pm$ 1.3	11.2 $\pm$ 1.3
DHA (mol% of fatty acid) *, mean $\pm$ SD	0.86 $\pm$ 0.78	0.88 $\pm$ 0.71

MUAC: mid upper arm circumference; Hb: hemoglobin; DHA: docosahexaenoic acid; * *n* = 258 (DHA); *n* = 224 (placebo).

The two groups did not differ in estimated intake of energy or any macronutrient at baseline (Supplementary Table S2 and Table 2). The mean DHA levels at baseline and delivery by birth weight (<2500 g; $\geq$2500 g), length (<50 cm; $\geq$50 cm), and head circumference (<34 cm; $\geq$34 cm) are shown in Table 3. The mean DHA levels at baseline did not differ overall between the two groups. A significant change was observed in the mean DHA values at delivery, being higher in the DHA group as compared to placebo, both overall and when subdivided by birth weight, length, head circumference, and gestational age.

Table 2. Mean DHA (mol% of fatty acid) levels in RBC phospholipids.

DHA Levels	DHA	Placebo	Mean Difference * (95% CI)	*p*-Value
	n, Mean $\pm$ SD, Median (p25, p75)	*n*, Mean $\pm$ SD, Median (p25, p75)		
Overall				
DHA at baseline	*n* = 256, 0.86 $\pm$ 0.78, 0.56 (0.31, 1.20)	*n* = 224, 0.88 $\pm$ 0.71, 0.55 (0.37, 1.28)	0.02 (−0.11, 0.15)	0.770
DHA at delivery	*n* = 269, 2.03 $\pm$ 1.76, 1.41 (0.61, 2.99)	*n* = 242, 1.12 $\pm$ 0.86, 0.83 (0.42, 1.72)	−0.91 (−1.16, −0.67)	<0.001
Birth Weight < 2500 g				
DHA at baseline	*n* = 63, 0.96 $\pm$ 0.89, 0.59 (0.39, 1.41)	*n* = 47, 0.74 $\pm$ 0.69, 0.46 (0.37, 0.95)	−0.22 (−0.53, 0.09)	0.170
DHA at delivery	*n* = 67, 2.00 $\pm$ 1.81, 1.39 (0.63, 2.79)	*n* = 50, 1.17 $\pm$ 0.80, 0.96 (0.48, 1.74)	−0.83 (−1.37, −0.29)	0.003
Birth Weight $\geq$ 2500 g				
DHA at baseline	*n* = 193, 0.83 $\pm$ 0.75, 0.53 (0.3, 1.11)	*n* = 177, 0.92 $\pm$ 0.71, 0.59 (0.37, 1.33)	0.09 (−0.06, 0.24)	0.221
DHA at delivery	*n* = 202, 2.04 $\pm$ 1.74, 1.43 (0.6, 3.17)	*n* = 192, 1.10$\pm$ 0.88, 0.78 (0.4, 1.7)	−0.94 (−1.22, −0.66)	<0.001
Gestation Age < 37 Weeks				
DHA at baseline	*n* = 18, 1.1 $\pm$ 0.74, 0.79 (0.61, 1.55)	*n* = 19, 0.55 $\pm$ 0.37, 0.41 (0.37, 0.66)	−0.55 (−0.93, −0.16)	0.007
DHA at delivery	*n* = 17, 2.24 $\pm$ 1.81, 1.72 (0.96, 2.97)	*n* = 19, 0.99 $\pm$ 0.72, 0.62 (0.41, 1.72)	1.25 (−2.16, −0.33)	0.009
Gestation Age $\geq$ 37 Weeks				
DHA at baseline	*n* = 238, 0.84 $\pm$ 0.79, 0.53 (0.31, 1.14)	*n* = 205, 0.91 $\pm$ 0.72, 0.59 (0.37, 1.3)	−0.07 (−0.07, 0.21)	0.333
DHA at delivery	*n* = 252, 2.02 $\pm$ 1.76, 1.4 (0.59, 3.08)	*n* = 223, 1.13 $\pm$ 0.87, 0.83 (0.42, 1.72)	−0.89 (−1.15, −0.64)	<0.001
Birth Length < 50 cm				
DHA at baseline	*n* = 234, 0.89 $\pm$ 0.81, 0.58 (0.32, 1.33)	*n* = 194, 0.85 $\pm$ 0.71, 0.52 (0.34, 1.23)	−0.04 (0.1, −0.33)	0.556
DHA at delivery	*n* = 245, 2.04 $\pm$ 1.77, 1.39 (0.62, 3.17)	*n* = 213, 1.1 $\pm$ 0.86, 0.78 (0.42, 1.72)	−0.94 (-0.68, −0.33)	<0.0001
Birth Length $\geq$ 50 cm				
DHA at baseline	*n* = 21, 0.51 $\pm$ 0.33, 0.44 (0.28, 0.59)	*n* = 30, 1.11 $\pm$ 0.64, 1 (0.62, 1.47)	0.6 (0.91, −0.33)	0.0003
DHA at delivery	*n* = 24, 1.97 $\pm$ 1.7, 1.72 (0.58, 2.45)	*n* = 29, 1.28 $\pm$ 0.87, 1.14 (0.54, 1.73)	−0.69 (0.04, −0.33)	0.063
Head Circumference < 34 cm				
DHA at baseline	N = 123, 0.95 $\pm$ 0.89, 0.59 (0.34, 1.36)	N = 98, 0.82 $\pm$ 0.66, 0.53 (0.39, 1.23)	−0.12 (0.09, −0.33)	0.258
DHA at delivery	n=132, 2.16 $\pm$ 1.81, 1.49 (0.73, 2.98)	N = 106, 1.16 $\pm$ 0.89, 0.93 (0.45, 1.74)	−1.00 (−0.62, −0.33)	<0.0001
Head Circumference $\geq$ 34 cm				
DHA at baseline	*n* = 132, 0.78 $\pm$ 0.67, 0.52 (0.29, 0.97)	*n* = 126, 0.93 $\pm$ 0.74, 0.57 (0.34, 1.4)	0.15 (0.32, −0.33)	0.0916
DHA at delivery	*n* = 137, 1.91 $\pm$ 1.7, 1.19 (0.55, 3.17)	*n* = 136, 1.08 $\pm$ 0.84, 0.78 (0.41, 1.7)	−0.83 (−0.51, −0.33)	<0.0001

DHA levels were analyzed only in a subset of women; data are presented as mean $\pm$ standard deviation, median (p25, p75); DHA: docosahexaenoic acid; *p*-Value calculated using unpaired *t*-test; * difference = placebo minus DHA.

Table 3. Birth outcomes for all live births according to treatment group.

Birth Outcomes	DHA		Placebo		Mean Difference [§] (95% CI)	*p*-Value
	n	Mean ± SD/n (%)	*n*	Mean ± SD/n (%)		
Gestational age at delivery (weeks)	440	38.8 ± 1.7	440	38.8 ± 1.7	0.07 (−0.16, 0.30)	0.54
Preterm birth (gestation < 37 week) [†]	440	28 (6.4%)	440	33 (7.5%)	0.01 (−0.02, 0.04)[‡]	0.52
Newborn Anthropometry						
Birth weight (grams)	440	2750.6 ± 421.5	440	2768.2 ± 436.6	17.6 (−39.2, 74.4)	0.54
Low birth weight (<2500 g) [†]	440	105 (23.9%)	440	99 (22.5%)	−0.01 (−0.07, 0.04) [‡]	0.63
Birth length (cm)	413	47.3 ± 2.0	410	47.5 ± 2.0	0.21 (−0.06, 0.48)	0.13
Birth head circumference (cm)	413	33.7 ± 1.4	410	33.8 ± 1.4	0.14 (−0.05, 0.34)	0.15
Apgar score at 1 min	372	6.9 ± 0.8	376	6.9 ± 0.8	0.01 (−0.11, 0.13)	0.91
Apgar score at 5 min	373	8.0 ± 0.7	378	8.0 ± 0.7	0.03 (−0.07, 0.12)	0.60
Size for Gestational Age and Sex According to Standardized Measures [¶]						
Birth weight for gestational age z score	440	−0.97 ± 0.98	440	−0.95 ± 0.95	0.03 (−0.1, 0.16)	0.67
Birth length for gestational age z score	413	−0.84 ± 1.04	410	−0.73 ± 1.12	0.11 (−0.03, 0.26)	0.13
Birth head circumference for gestational age z score.	413	0.09 ± 1.05	410	0.20 ± 0.97	0.11 (−0.03, 0.25)	0.11
Small for gestational age *,[@]	440	172 (39.1%)	440	172 (39.1%)	na	na

[†] *n* (%); [‡] difference in proportions reported; [§] difference = (placebo − DHA); difference in mean values reported using two-sample *t*-test. Difference in proportions reported using proportion test; [¶] standards are based on those established by the INTERGROWTH-21st (International Fetal and Newborn Growth Consortium for the 21st century) Project [23]. [@]: Infants considered to be small for gestational age had a weight-for-age z score that was below the 10th percentile according to neonatal standards established by the INTERGROWTH-21st Project. Na: not applicable.

Supplementary Materials Table S3 shows the mean change in DHA levels from baseline to delivery by treatment group. There is an increase in mean DHA level in both the groups from baseline to delivery (DHA 1.20 (0.98, 1.43), *p* = <0.001; placebo 0.24 (0.11, 0.36), *p* = 0.0002).

There were 450 (94.1%) and 452 (94.3%) live births in the DHA (*n* = 478) and placebo (*n* = 479) groups, respectively. Compliance was high in both groups (DHA: 88.5% and placebo: 87.1%). There were 230 (52.3%) and 235 (53.4%) male children in the DHA and placebo groups, respectively, and percentages were calculated based on 440 (DHA) and 440 (placebo) analyzed neonates.

3.2. Outcomes

Table 3 shows birth outcomes for all live births according to the treatment group. There were no significant differences between DHA and placebo groups for mean birth weight (2750.6 ± 421.5 vs. 2768.2 ± 436.6 g, *p* = 0.54), birth length (47.3 ± 2.0 vs. 47.5 ± 2.0 cm, *p* = 0.13), or head circumference (33.7 ± 1.4 vs. 33.8 ± 1.4 cm, *p* = 0.15). The APGAR scores at 1 min and 5 min were similar between the groups. We did not find any significant difference between DHA and placebo groups in z scores for birth weight, length, and head circumference.

Gestational age at delivery was similar between DHA and placebo groups (DHA vs. placebo: 38.8 ±1.7 vs. 38.8 ± 1.7 wk, *p* = 0.54). The prevalence of preterm birth and low birth weight did not differ significantly between the groups. Unfortunately, the causes of preterm birth and the number of intra uterine growth retardation (IUGR) cases were not collected.

3.3. Sub-Group Analysis

Figures 2–4 show the results of sub-group analyses for birth weight, birth length, and head circumference, respectively. The effect of DHA on the birth size (i.e., weight, length, and head circumference) did not differ across any of the subgroups examined (*p* = 0.007, *p*-value adjusted for multiplicity using Bonferroni correction). Similarly, there was no evidence of differences by compliance, the gender of the child, or preterm status.

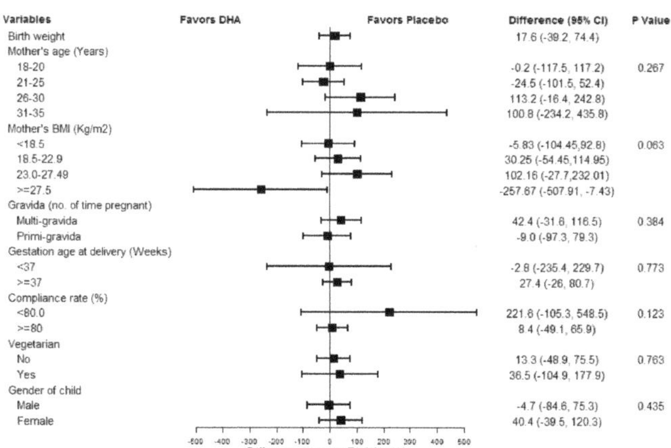

Figure 2. Sub-group analysis for newborn birthweight.

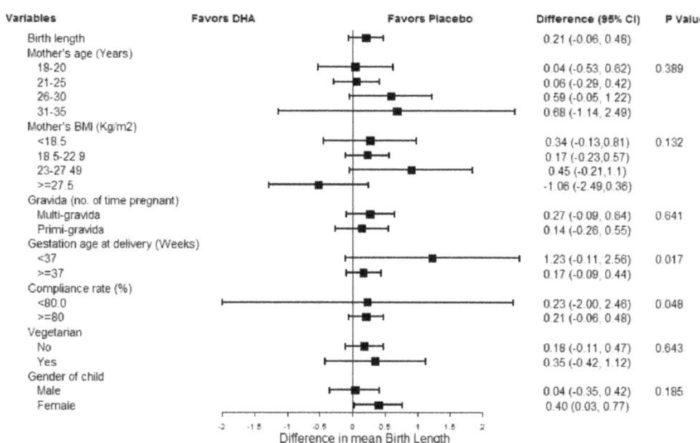

Figure 3. Sub-group analysis for newborn birth length.

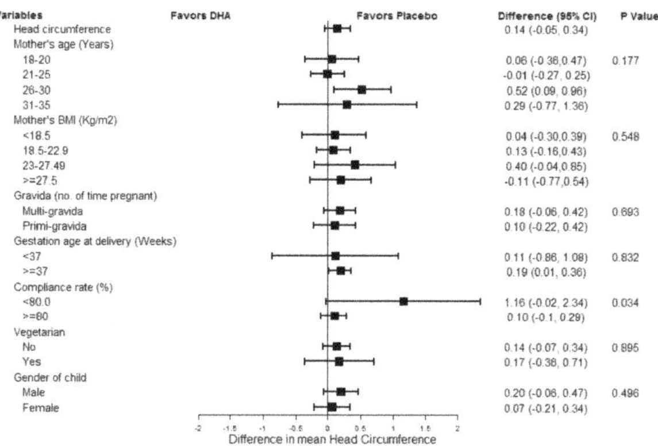

Figure 4. Sub-group analysis for newborn head circumference.

4. Discussion

In this study, maternal supplementation with 400 mg/day DHA in the second half of pregnancy did not affect the weight, length, or head circumference of the offspring at birth.

While this was in contrast to findings from some high-income settings [14], it concurs with other studies from relatively comparable settings [21].

Although mechanistic pathways linking maternal polyunsaturated fatty acid (PUFA), especially DHA status with gestational length, are poorly delineated, prenatal DHA supplementation has been shown to enhance the gestation duration in some studies [26]. This longer gestation duration with fish oil that contains EPA as well as DHA may be due to an alteration in the balance of prostaglandins derived from EPA and arachidonic acid [27]. A high proportion of omega-6 to omega-3 FAs can contribute to increased pro-inflammatory eicosanoids (i.e., prostaglandin E2 (PGE2) and prostaglandin F2 (PGF2)) production. These metabolites have been shown to be linked with the initiation of labor and premature labor. Including more EPA in the diet may lead to a reduction in the production of pro-inflammatory eicosanoids and expanded production of prostacyclin (PGI2), which may promote myometrial relaxation. Omega-3 LC-PUFA, especially DHA, downregulates production of prostaglandins PGE2 and PGF2 and may thus inhibit the process of parturition. This has been postulated to be associated with increased gestation duration and the accretion of intrauterine LC-PUFA [28]. Longer gestation indeed also influences newborn anthropometry positively, and thus DHA was shown to also confer small benefits on newborn anthropometry because of its impact on gestation duration. However, our trial did not find any such benefit.

A recent 2018 Cochrane review [29] looking at the impact of omega 3 fatty acids (including both DHA and EPA) concludes that omega 3 LCPUFA reduces the incidence of preterm birth <37 weeks and early preterm birth <34 weeks in women receiving omega-3 LCPUFA compared with no omega-3s. Thus, in our study the supplementation of DHA without other fractions like EPA may not have been able to result in an effect on gestation length. This review of high-quality evidence from 15 trials with 8449 participants also noted that there was a reduced risk LBW (15.6% versus 14%; RR 0.90, 95% CI 0.82 to 0.99) [29]. Increased birthweight due to prenatal DHA supplementation has been observed in only primiparous women [30]. The authors suggest that since primiparous women were, on average, younger than multiparous women, their own body stores of DHA are not well established and available to the fetus and infant [30]. Ramakrishnan et al., from the same cohort, showed that the offspring of primigravid women who received DHA were heavier at birth than the offspring of primigravid women who received placebo (difference, 99.4 g; 95% CI, 5.5 to 193.4) and had larger head circumferences (difference, 0.5 cm; 95% CI, 0.1 to 0.9 cm) [21]. In the current study, however, the woman's parity did not affect the effect of DHA on the newborn's birth weight, length, or head circumference.

Key strengths of this study are the strong study design combined with high retention rates and compliance (verified by the rise in erythrocyte DHA levels).

Another parameter which is often of interest is the timing of initiation of supplementation during pregnancy. The other salient trials [14,19] initiated DHA supplementation during mid pregnancy (14.5 weeks and 19 weeks median, respectively). Similarly, in our trial, DHA supplementation started between 12 and 18 weeks of pregnancy, with a median value of 15 weeks. Nevertheless, we did not observe any impact of DHA on the outcomes, unlike the two other trials.

The complexity of multiple other factors apart from DHA in affecting birth size needs to be recognized. Factors like the maternal diet at multiple time points during pregnancy, family support, stress levels [31], and the consumption of other important micronutrients like iron and zinc that were not assessed may have influenced birth size [32]. Further, we do not have data on the single nucleotide polymorphisms (SNPs) in the fatty acid desaturase (FADS) gene that has been known to affect the activity of the enzymes that convert PUFAs into their long-chain active form and may determine who benefits from supplementation [33,34]. Future large-scale trials taking into account all these factors are warranted.

5. Conclusions

In summary, no beneficial effects of prenatal supplementation of Indian women with DHA from mid-pregnancy through delivery on newborn anthropometry were observed.

Supplementary Materials: The following are available online at https://www.mdpi.com/2072-6643/13/3/730/s1: Table S1: Comparison of baseline characteristics comparing women who continued to participate in the study through delivery and those who did not, Table S2: Dietary data on subsample at randomization (n = 278); Table S3: Mean change in DHA levels from baseline to delivery.

Author Contributions: S.K. had full access to all the data in the study and takes responsibility for the integrity of the data and the accuracy of the data analysis. She also takes the final responsibility for the decision to submit for publication. Conceptualization, S.K., D.P., N.T., U.R., and A.D.S.; Data curation, D.K.; Formal analysis, D.K. and R.G.; Funding acquisition, S.K. and D.P.; Investigation, S.K., G.P., S.B.M., and Y.K.; Methodology, S.K. and U.R.; Project administration, S.K., M.C., and K.P.; Resources, K.P., M.K.S., and D.P.; Software, D.K.; Supervision, K.P., M.K.S., N.T., and A.D.S.; Validation, R.G.; Writing—original draft, S.K.; Writing—review and editing, D.K., M.C., D.P., N.T., U.R., and A.D.S. All authors have read and agreed to the published version of the manuscript.

Funding: The trial was funded by India Alliance IA/CPHE/14/1/501498. The DST Young Scientist Award (SR/FT/LS-156/2011) also provided partial funding for setting up a part of the cohort. The supplements were donated by DSM Nutritional Products via their Mumbai office.

Institutional Review Board Statement: The study was conducted according to the guidelines of the Declaration of Helsinki, and approved by the Institutional Review Board (or Ethics Committee) of Center for Chronic Disease Control (CCDC-IEC_04_2015), Public Health Foundation of India (PHFI) (TRC-IEC-261/15), and Jawaharlal Nehru Medical College (MDC/IECHSR/2016-17/A-85).

Informed Consent Statement: Informed consent was obtained from all subjects involved in the study.

Data Availability Statement: The request for accessing de-identified data plus data dictionary will be put forth for approval by the trial's mentoring and advisory committee (MAC). Interested researchers may submit a proposal with a valid reason or justification, e.g., meta-analysis, etc.

Acknowledgments: The Department of Obstetrics and Gynecology at KAHER's JN Medical College (JNMC) Belagavi is deeply acknowledged. The Unit Chiefs (M.B. Bellad, Anita Dalal, Yeshita Pujar, and M.C. Metgud) deserve a special mention in providing valuable inputs and support during each phase of this study. We are indebted to the JNMC's Blood Bank in the charge of Shrikant Viragi and his whole team for facilitating all on site biochemical work in the study. We are truly grateful to the resource and support teams at the Centre for Chronic Disease Control (CCDC) and Public Health Foundation of India (PHFI), Delhi. This trial is funded by Wellcome Trust-DBT India Alliance (December 2015–December 2020). The Young Scientist Award by DST SERB India (2013–16) helped us establish the DHANI trial. The funding sources had no role in the design and conduct of the study; collection, management, analysis, and interpretation of the data; preparation, review, or approval of the manuscript; and the decision to submit the manuscript for publication. Neither the product provider nor the sponsors had an opportunity to review a pre-submission copy of the article.

Conflicts of Interest: The authors declare no conflict of interest. The funders had no role in the design of the study; in the collection, analyses, or interpretation of data; in the writing of the manuscript; or in the decision to publish the results.

References

1. Law, C.M. Significance of birth weight for the future. *Arch. Dis. Child. -Fetal Neonatal Ed.* **2002**, *86*, F7. [CrossRef]
2. Blencowe, H.; Krasevec, J.; De Onis, M.; Black, R.E.; An, X.; Stevens, G.A.; Borghi, E.; Hayashi, C.; Estevez, D.; Cegolon, L.; et al. National, regional, and worldwide estimates of low birthweight in 2015, with trends from 2000: A systematic analysis. *Lancet Glob. Health* **2019**, *7*, e849–e860. [CrossRef]
3. Lee, A.C.C.; Katz, J.; Blencowe, H.; Cousens, S.; Kozuki, N.; Vogel, J.P.; Adair, L.; Baqui, A.H.; Bhutta, Z.A.; Caulfield, L.E.; et al. National and regional estimates of term and preterm babies born small for gestational age in 138 low-income and middle-income countries in 2010. *Lancet Glob. Health* **2013**, *1*, e26–e36. [CrossRef]
4. Nordman, H.; Jääskeläinen, J.; Voutilainen, R. Birth Size as a Determinant of Cardiometabolic Risk Factors in Children. *Horm. Res. Paediatr.* **2020**, *93*, 144–153. [CrossRef]

5.   Groer, M.W.; Gregory, K.E.; Louis-Jacques, A.; Thibeau, S.; Walker, W.A. The very low birth weight infant microbiome and childhood health. *Birth Defects Res. Part C Embryo Today Rev.* **2015**, *105*, 252–264. [CrossRef]

6.   Kanda, T.; Murai-Takeda, A.; Kawabe, H.; Itoh, H. Low birth weight trends: Possible impacts on the prevalences of hypertension and chronic kidney disease. *Hypertens. Res.* **2020**, *43*, 859–868. [CrossRef] [PubMed]

7.   Brown, H.L.; Smith, G.N. Pregnancy Complications, Cardiovascular Risk Factors, and Future Heart Disease. *Obstet. Gynecol. Clin. N. Am.* **2020**, *47*, 487–495. [CrossRef]

8.   Yajnik, C.S.; Deshmukh, U.S. Maternal nutrition, intrauterine programming and consequential risks in the offspring. *Rev. Endocr. Metab. Disord.* **2008**, *9*, 203–211. [CrossRef]

9.   Stein, A.D.; Barros, F.C.; Bhargava, S.K.; Hao, W.; Horta, B.L.; Lee, N.; Kuzawa, C.W.; Martorell, R.; Ramji, S.; Stein, A.; et al. Birth Status, Child Growth, and Adult Outcomes in Low- and Middle-Income Countries. *J. Pediatr.* **2013**, *163*, 1740–1746.e4. [CrossRef]

10. Cutland, C.L.; Lackritz, E.M.; Mallett-Moore, T.; Bardají, A.; Chandrasekaran, R.; Lahariya, C.; Nisar, M.I.; Tapia, M.D.; Pathirana, J.; Kochhar, S.; et al. Low birth weight: Case definition & guidelines for data collection, analysis, and presentation of maternal immunization safety data. *Vaccine* **2017**, *35*, 6492–6500. [CrossRef]

11. Kader, M.; Perera, N.K.P.P. Socio-economic and nutritional determinants of low birth weight in India. *N. Am. J. Med Sci.* **2014**, *6*, 302–308. [CrossRef]

12. Das, J.K.; Hoodbhoy, Z.; Salam, R.A.; Bhutta, A.Z.; Valenzuela-Rubio, N.G.; Prinzo, Z.W.; Bhutta, Z.A. Lipid-based nutrient supplements for maternal, birth, and infant developmental outcomes. *Cochrane Database Syst. Rev.* **2018**, *8*, CD012610. [CrossRef]

13. Velzing-Aarts, F.; Van Der Klis, F.; Van Der Dijs, F.; Van Beusekom, C.; Landman, H.; Capello, J.; Muskiet, F. Effect of three low-dose fish oil supplements, administered during pregnancy, on neonatal long-chain polyunsaturated fatty acid status at birth. *Prostaglandins Leukot. Essent. Fat. Acids* **2001**, *65*, 51–57. [CrossRef]

14. Carlson, E.S. Docosahexaenoic acid supplementation in pregnancy and lactation. *Am. J. Clin. Nutr.* **2008**, *89*, 678S–684S. [CrossRef]

15. Tupe, R.; Chiplonkar, S.A. Diet patterns of lactovegetarian adolescent girls: Need for devising recipes with high zinc bioavailability. *Nutrition* **2010**, *26*, 390–398. [CrossRef] [PubMed]

16. Dwarkanath, P.; Muthayya, S.; Thomas, T.; Vaz, M.; Parikh, P.; Mehra, R.; Kurpad, A.V. Polyunsaturated fatty acid consumption and concentration among South Indian women during pregnancy. *Asia Pac. J. Clin. Nutr.* **2009**, *18*, 389–394. [PubMed]

17. Food and Agriculture Organization (FAO). *Fats and Fatty Acids in Human Nutrition: Report of an Expert Consultation*; FAO: Rome, Italy, 2010; FAO Food and Nutrition Paper 91.

18. Muthayya, S.; Dwarkanath, P.; Thomas, T.; Ramprakash, S.; Mehra, R.; Mhaskar, A.; Mhaskar, R.; Thomas, A.; Bhat, S.; Vaz, M.; et al. The effect of fish and omega-3 LCPUFA intake on low birth weight in Indian pregnant women. *Eur. J. Clin. Nutr.* **2009**, *63*, 340–346. [CrossRef]

19. Makrides, M.; Best, K. Docosahexaenoic Acid and Preterm Birth. *Ann. Nutr. Metab.* **2016**, *69*, 29–34. [CrossRef]

20. Carlson, E.S.; Colombo, J.; Gajewski, B.J.; Gustafson, K.M.; Mundy, D.; Yeast, J.; Georgieff, M.K.; Markley, L.A.; Kerling, E.H.; Shaddy, D.J. DHA supplementation and pregnancy outcomes. *Am. J. Clin. Nutr.* **2013**, *97*, 808–815. [CrossRef]

21. Ramakrishnan, U.; Stein, A.D.; Parra-Cabrera, S.; Wang, M.; Imhoff-Kunsch, B.; Juárez-Márquez, S.; Rivera, J.; Martorell, R. Effects of Docosahexaenoic Acid Supplementation During Pregnancy on Gestational Age and Size at Birth: Randomized, Double-Blind, Placebo-Controlled Trial in Mexico. *Food Nutr. Bull.* **2010**, *31*, S108–S116. [CrossRef]

22. Makrides, M.; Gibson, R.A.; McPhee, A.J.; Yell, L.; Quinlivan, J.; Ryan, P. Effect of DHA Supplementation During Pregnancy on Maternal Depression and Neurodevelopment of Young Children: A Randomized Controlled Trial. *Obstet. Gynecol. Surv.* **2011**, *66*, 79–81. [CrossRef]

23. Khandelwal, S.; Kondal, D.; Chaudhry, M.; Patil, K.; Swamy, M.K.; Metgud, D.; Jogalekar, S.; Kamate, M.; Divan, G.; Gupta, R.; et al. Effect of Maternal Docosahexaenoic Acid (DHA) Supplementation on Offspring Neurodevelopment at 12 Months in India: A Randomized Controlled Trial. *Nutrients* **2020**, *12*, 3041. [CrossRef]

24. Villar, J.; Ismail, L.C.; Victora, C.G.; Ohuma, E.O.; Bertino, E.; Altman, D.G.; Lambert, A.; Papageorghiou, A.T.; Carvalho, M.; Jaffer, Y.A.; et al. International standards for newborn weight, length, and head circumference by gestational age and sex: The Newborn Cross-Sectional Study of the INTERGROWTH-21st Project. *Lancet* **2014**, *384*, 857–868. [CrossRef]

25. Wold Health Organization. WHO expert consultation, Appropriate body-mass index for Asian populations and its implications for policy and intervention strategies. *Lancet* **2004**, *363*, 157–163. [CrossRef]

26. Harris, M.A.; Reece, M.S.; McGregor, J.A.; Wilson, J.W.; Burke, S.M.; Wheeler, M.; Anderson, J.E.; Auld, G.W.; French, J.I.; Allen, K.G.D. The Effect of Omega-3 Docosahexaenoic Acid Supplementation on Gestational Length: Randomized Trial of Supplementation Compared to Nutrition Education for Increasing n-3 Intake from Foods. *BioMed Res. Int.* **2015**, *2015*, 123078. [CrossRef]

27. Calder, P.C. Docosahexaenoic Acid. *Ann. Nutr. Metab.* **2016**, *69* (Suppl. 1), 7–21. [CrossRef]

28. Allen, K.G.; Harris, M.A. The Role of n-3 Fatty Acids in Gestation and Parturition. *Exp. Biol. Med.* **2001**, *226*, 498–506. [CrossRef] [PubMed]

29. Middleton, P.; Gomersall, J.C.; Gould, J.F.; Shepherd, E.; Olsen, S.F.; Makrides, M. Omega-3 fatty acid addition during pregnancy. *Cochrane Database Syst. Rev.* **2018**, *11*, CD003402. [CrossRef]

30. Stein, A.D.; Wang, M.; Martorell, R.; Neufeld, L.M.; Flores-Ayala, R.; Rivera, J.A.; Ramakrishnan, U. Growth to Age 18 Months Following Prenatal Supplementation with Docosahexaenoic Acid Differs by Maternal Gravidity in Mexico. *J. Nutr.* **2010**, *141*, 316–320. [CrossRef]

31. Varma, J.R.; Nimbalkar, S.M.; Patel, D.; Phatak, A.G. The Level and Sources of Stress in Mothers of Infants Admitted in Neonatal Intensive Care Unit. *Indian J. Psychol. Med.* **2019**, *41*, 338–342. [CrossRef]
32. Mousa, A.; Naqash, A.; Lim, S. Macronutrient and Micronutrient Intake during Pregnancy: An Overview of Recent Evidence. *Nutrients* **2019**, *11*, 443. [CrossRef]
33. Gonzalez-Casanova, I.; Rzehak, P.; Stein, A.D.; Feregrino, R.G.; Dommarco, J.A.R.; Barraza-Villarreal, A.; Demmelmair, H.; Romieu, I.; Villalpando, S.; Martorell, R.; et al. Maternal single nucleotide polymorphisms in the fatty acid desaturase 1 and 2 coding regions modify the impact of prenatal supplementation with DHA on birth weight. *Am. J. Clin. Nutr.* **2016**, *103*, 1171–1178. [CrossRef] [PubMed]
34. Scholtz, S.; Kerling, E.; Shaddy, D.; Li, S.; Thodosoff, J.; Colombo, J.; Carlson, S. Docosahexaenoic acid (DHA) supplementation in pregnancy differentially modulates arachidonic acid and DHA status across FADS genotypes in pregnancy. *Prostaglandins Leukot. Essent. Fat. Acids* **2015**, *94*, 29–33. [CrossRef]

Article

Effect of Maternal Docosahexaenoic Acid (DHA) Supplementation on Offspring Neurodevelopment at 12 Months in India: A Randomized Controlled Trial

Shweta Khandelwal [1,2,*], Dimple Kondal [1,2], Monica Chaudhry [1], Kamal Patil [3],
Mallaiah Kenchaveeraiah Swamy [3], Deepa Metgud [4], Sandesh Jogalekar [3], Mahesh Kamate [3],
Gauri Divan [5,6], Ruby Gupta [1,2], Dorairaj Prabhakaran [1,2], Nikhil Tandon [7],
Usha Ramakrishnan [8] and Aryeh D. Stein [8]

[1] Public Health Foundation of India, 47, Sector 44, Institutional area, Gurugram, Haryana 122003, India; dimple@ccdcindia.org (D.K.); monica.chaudhry@phfi.org (M.C.); ruby.gupta@phfi.org (R.G.); dprabhakaran@phfi.org (D.P.)
[2] Centre for Chronic Disease Control, C-1/52, 2nd Floor, Safdarjung Development Area, New Delhi 110016, India
[3] KAHER's JN Medical College, JNMC KLE University Campus, Nehru Nagar, Belgaum, Karnataka 590010, India; kamalpatil1967@yahoo.co.in (K.P.); mkswamy53@yahoo.co.in (M.K.S.); sandeshjogalekar555@gmail.com (S.J.); drmaheshkamate@gmail.com (M.K.)
[4] KAHER's Institute of Physiotherapy, JNMC KLE University Campus, Nehru Nagar, Belgaum, Karnataka 590010, India; drdeepa_metgud@yahoo.com
[5] Sangath, C-1/52, Block C 1, Bhim Nagri, Hauz Khas, New Delhi 110016, India; gauri.divan@sangath.in
[6] Sangath Goa, H No 451 (168), Bhatkar Waddo, Socorro, Porvorium, Bardez, Goa 403501, India
[7] All India Institute of Medical Sciences, Sri Aurobindo Marg, New Delhi 110029, India; nikhil_tandon@hotmail.com
[8] Hubert Department of Global Health, Rollins School of Public Health, Emory University, Atlanta, GA 30322, USA; uramakr@emory.edu (U.R.); aryeh.stein@emory.edu (A.D.S.)
* Correspondence: shweta.khandelwal@phfi.org; Tel.: +91-0-9891111858

Received: 14 August 2020; Accepted: 23 September 2020; Published: 3 October 2020

Abstract: Intake of dietary docosahexaenoic acid (DHA 22:6n-3) is very low among Indian pregnant women. Maternal supplementation during pregnancy and lactation may benefit offspring neurodevelopment. We conducted a double-blind, randomized, placebo-controlled trial to test the effectiveness of supplementing pregnant Indian women (singleton gestation) from ≤20 weeks through 6 months postpartum with 400 mg/d algal DHA compared to placebo on neurodevelopment of their offspring at 12 months. Of 3379 women screened, 1131 were found eligible; 957 were randomized. The primary outcome was infant neurodevelopment at 12 months, assessed using the Development Assessment Scale for Indian Infants (DASII). Both groups were well balanced on sociodemographic variables at baseline. More than 72% of women took >90% of their assigned treatment. Twenty-five serious adverse events (SAEs), none related to the intervention, (DHA group = 16; placebo = 9) were noted. Of 902 live births, 878 were followed up to 12 months; the DASII was administered to 863 infants. At 12 months, the mean development quotient (DQ) scores in the DHA and placebo groups were not statistically significant (96.6 ± 12.2 vs. 97.1 ± 13.0, $p = 0.60$). Supplementing mothers through pregnancy and lactation with 400 mg/d DHA did not impact offspring neurodevelopment at 12 months of age in this setting.

Keywords: maternal supplementation; pregnancy; lactation; docosahexaenoic acid (DHA); neurodevelopment; randomized controlled trial (RCT); India

1. Introduction

The first 1000 days are crucial for a child's neurodevelopment [1]. The brain develops rapidly through neurogenesis, axonal and dendritic growth, synaptogenesis, synaptic pruning, myelination, and gliogenesis [2]. These events build on each other, such that even small perturbations can have long-term effects on the brain's structural and functional capacity [3]. Maternal nutrition during this time influences both pre- and postnatal growth, and development of the offspring [4–6].

It has been suggested that n-3 long-chain polyunsaturated fatty acids (LCPUFA) (especially docosahexaenoic acid [DHA]) levels enhance infant neurodevelopment [7–11]. The DHA is an important structural component of the human brain and retina. DHA accumulates in all of the brain regions and retinal photoreceptors [12]. These long-chain fatty acids regulate the fluidity of cell membranes as well as the activity of ion channels, enabling synaptic transmission and providing substrate binding to membrane receptors. The deposition of DHA in human brain phospholipids occurs primarily during the last trimester of pregnancy such as week 30 until the early postnatal periods continuing during the first two years of life [13–15]. Human fetuses and young infants have limited ability to synthesize n-3 LCPUFA de novo and are supplied via maternal (placental transfer, breast milk) or external (formula, dietary) sources [16,17]. Approximately 67 mg of DHA is accrued by the fetus per day [18]. Deprivation of n-3 LCPUFA, whether prenatally or after birth, has deleterious effects on learning abilities, memory, and visual grating acuity in monkeys, rats, and human infants [19,20].

DHA can be obtained from marine algae, fatty fish, and marine oils. Endogenous synthesis of DHA is limited, especially in the presence of excess n-6 LCPUFA. The Food and Agriculture Organization and World Health Organization (FAO/WHO) Expert Committee recommends a diet with a 5–10:1 ratio of n-6/n-3 LCPUFA and 300 mg/day of preformed DHA during pregnancy [21,22]. Since cereal-based diets are rich in n-6 LCPUFA but largely deficient in DHA-rich sources, population levels of plasma DHA are low in India. Studies report [23,24] that mean DHA intake was lowest (11 mg) among Indian pregnant women in the third trimester [25] compared to pregnant women from other developing countries like Bangladesh, Burkina Faso, Chile, China, India, and Mexico.

Studies conducted around the world to assess the association between intakes of DHA during pregnancy or lactation and neurodevelopmental outcomes in childhood have been inconsistent [7,26–35]. Few studies have assessed the effect of both prenatal and postnatal intake of DHA [36,37]. There is a paucity of data on the potential benefits of maternal DHA supplementation in infants, especially in the Indian population. The present study examines the hypothesis whether 400 mg/d maternal DHA supplementation from ≤20 weeks through 6 months postpartum influences infant neurodevelopment at 12 months of age. The present study DHANI (Maternal **DHA** supplementation and offspring Neurodevelopment in India) is the first to examine the effects of maternal DHA supplementation from mid-pregnancy through six months postpartum on postnatal neurodevelopment in India.

2. Materials and Methods

2.1. Study Design

DHANI was a randomized, double-blinded, placebo-controlled trial to test the effect of providing pregnant women 400 mg/d algal DHA compared to placebo from ≤20 weeks of singleton gestation through 6 months postpartum on offspring neurodevelopment. The trial protocol has been published elsewhere [38]. The protocol was reviewed and approved by institutional review boards (IRBs) of all participating institutions: Center for Chronic Disease Control (CCDC-IEC_04_2015), Public Health Foundation of India (TRC-IEC-261/15), Jawaharlal Nehru Medical College (MDC/IECHSR/2016-17/A-85), and All India Institute of Medical Sciences (IEC-28/17.11.2015).

2.2. Participants

Healthy pregnant women (18–35 years; ≤20 weeks single gestation; with no medical complications or chronic diseases) attending the Department of Obstetrics and Gynecology at the Prabhakar Kore hospital (PKH) in a large city in southwest India for their prenatal check-ups were approached by designated project staff. The consulting obstetrician on site, considering obstetric history and complications, affirmed final eligibility. After signing a witnessed informed consent form in their preferred local language (Kannada, Marathi, or Hindi), consenting women (n = 957) were randomized to receive either 400 mg of DHA or a placebo until six months postpartum as explained later. Baseline assessments at enrollment included sociodemographic characteristics, dietary intake, obstetric history, anthropometric measurements, blood investigations including a non-fasting blood draw, vital signs. The women received their first 15-day supply of supplements in the form of coded bottles. Further supplements were provided at the women's homes every fortnight by fieldworkers.

The next study hospital visit was at delivery (referred to as E0). A research officer stayed in close contact with each mother after she entered her last trimester and visited the mother within 24 h of delivery to collect information on the type of delivery, complications if any, and newborn anthropometry, and obtained a breast milk sample. We also collected and stored maternal and cord blood samples from the enrolled mother–child dyads.

Postpartum visits for mother–child dyads at 1 month (E1), at 6 months (E6), and at 12 months (E12) postpartum were scheduled. E1 data collection (neonatal anthropometry, maternal anthropometry, breastfeeding pattern, breast milk sample (transported to site in ice box), overall health of both mother and child, etc.) was carried out at home by trained fieldworkers. Infant neurodevelopment was assessed at the hospital at E6 and E12 by trained and certified psychologists.

A pre-piloted and validated semi-quantitative food frequency questionnaire (FFQ) with n-3 LCPUFA-rich Indian foods was administered to all women at enrollment. Dietary data were also collected in a subsample of women (n = 278) using a 24-h diet recall within 1 month of recruitment. Nutrient intakes were calculated using DIETSOFT software (http://dietsoft.in/) based on standardized nutritive values of Indian foods [39]. A study physician reviewed adverse events.

2.3. Randomization and Masking

A computer-generated randomization using a permuted block design was used to generate the randomization list. The code list was generated for 1200 women to allow for potential loss to follow up. The assignment code list was placed in a sealed envelope at the beginning of the study and was held in a sealed location by a staff member at Gurugram, Public Health Foundation of India (PHFI). This list was used by this person to pre-code the supplement bottles in the warehouse before the bottles arrived on site, ready for distribution. All study participants and members of the study team (including members at the study site) remained blinded to the treatment allocation until the analysis of all data collected was carried out. Blinded preliminary descriptive analyses were presented to Data Safety Monitoring Board (DSMB) and once approved, full statistical analyses were undertaken. Unblinding occurred only after primary tables were generated.

The intervention was either 400 mg/d algal DHA or a matching placebo (soy/corn oil) delivered in the form of similar-looking softgel capsules, donated by DSM Nutritional Supplements, Mumbai. The composition of all capsules was same except for the DHA. Enrolled women received the intervention from the date of randomization until six months postpartum.

The shelf life of the capsules at room temperature (25 °C) was 24 months from the date of manufacture (90 days once the bottle was opened). The coded capsule bottles were stored in an on-site refrigerator for extra safety to slow down oxidation. Each bottle contained a fortnightly supply of capsules. Bottles were distributed by fieldworkers during scheduled home visits. The women were instructed to take two capsules (each with 200 mg DHA or placebo) daily, preferably at the same time each day. They were told to keep them in a cool, dry place. Supplements were provided for more than two weeks in cases where the woman shared plans to travel.

2.4. Procedures

Subjects were asked to maintain a daily record in an easy-to-fill log (piloted and found suitable for population with low literacy levels) of their supplement consumption. Weekly calls were made to encourage compliance and inquire about the general well-being of the women. The used bottles were collected (for pill count) by the fieldworkers during the fortnightly home visits. Red blood cell (RBC) phospholipids were analyzed for DHA levels. The analyses of breast milk samples have not been presented here.

2.5. Outcomes

Development quotient (DQ) as a marker of neurodevelopment, among infants at 12 months of age, was assessed by a trained psychologist using the Developmental Assessment Scale for Indian Infants (DASII). The DASII tool is the Indian modification of the Bayley Scales of Infant Development (BSID), using Indian norms for 67 motor and 163 mental items. DASII provides a measure of development for Indian infants below 30 months of age. DQ is defined as the ratio of functional to chronological age. Third, 50th and 97th percentile norms have been generated for Indian children (DQ range of 35–160 ± 3.5). The maximum DQ score is 160. In terms of interpretation of scores, DQ score ≥85 is normal and 71–84 is mild to moderate delay. Severe developmental delay is defined on DASII as DQ score ≤70 (≤2 SD). The inter-correlation between the motor and mental performance on the two sections of the scale ranges from 0.24 to 0.62 [40]. The motor development items cover the child's development from supine to erect posture, neck control, locomotion, and manipulative behavior such as reaching, picking up, handling things, and so forth. The mental development items record the child's cognizance of objects in the surroundings, perceptual pursuit of moving objects, development of communication and language comprehension, spatial relationship and manual dexterity, imitative behavior, social interaction, and so forth.

2.6. Biomarkers

Fatty acid composition of plasma phospholipids and red blood cell (RBC) membrane phospholipids are appropriate biomarkers of fatty acid status and related to dietary intakes [41,42]. Since the plasma (reflecting short-term intake) and RBC (reflecting long-term intake) fatty acid composition are related, the markers of longer-term intake RBC phospholipid fatty acids (linoleic acid (18:2n-6); alpha-linolenic acid (18:3n-3); arachidonic acid (20:4n-6); eicosapentanoic acid (20:5n-3); DHA (22:6n-3)) were measured. Lipids were extracted from RBCs using the method of Rose and Ocklander [43], phospholipids were separated by thin-layer chromatography. The phospholipids were trans-esterified using the method by Lepage and Roy [44] and fatty acids were identified by gas chromatography with flame ionization detector. Thirty-seven fatty acid methyl ester mix from Supelco (SIGMA-ALDRICH) was used to identify the fatty acids using their retention time.

Maternal nonfasting blood samples (5 mL) were obtained by venipuncture at recruitment, delivery, and six months postpartum. Neonatal blood samples were obtained from the umbilical cord vein immediately after delivery using the syringe method. A 2 mL venous blood sample was obtained from infants at 6 and 12 months of age. All samples were collected into tubes containing disodium ethylene diamine tetra acetic acid (EDTA). Plasma was separated the same day by cold centrifugation (4 °C) at 800× g for 10 min and RBCs were washed thrice using equal volumes of saline. Plasma and washed RBCs were stored at −80 °C for later analysis.

Breast milk samples (one day, one month, and six months postpartum) were collected. The milk samples were from a morning feed but not the first one, preferably between 08:00 am and 12:00 pm. Infants were allowed to suckle the nipple for a few minutes, and then a breast milk sample (10 mL) was expressed manually by the mother herself, after which the feeding continued. The samples were refrigerated immediately to prevent bacterial growth and later aliquoted into smaller 2 mL containers, filled nearly to the top to minimize oxidation, and frozen at −80 °C until analysis. Although the amount of fat may change within and between feeds, the proportion of fatty acid remains relatively constant. Since breast milk PUFA were being expressed as a percent of total fatty acids, complete breast expressions were not required [45]. RBC phospholipids were analyzed by gas chromatography using standard methods.

2.7. Statistical Analysis

The primary study outcome was infant development quotient (DQ) score (composite score of motor and mental development) obtained at 12 months of age. We required 674 mother–child pairs to detect an effect size of 0.25 with 90% power [46].

Maternal household and offspring characteristics in the two treatment groups were summarized to assess the effectiveness of randomization. Continuous, normally distributed variables were summarized as means and standard deviations, while skewed variables were summarized as medians and interquartile ranges. Categorical/binary variables were summarized using proportions. Analysis was done using the intent to treat (ITT) principle. Per protocol analyses were also carried out including infants who had a valid 12-month DQ score (DASII test administered at 12 months postpartum ± 4 weeks) and whose mothers' compliance rate >80%. All protocol deviation cases were excluded from per protocol analyses. We compared baseline characteristics between the final study sample and those who were lost to follow-up.

For the primary outcome, we used a two-sample t test to compare the mean DQ (mental and motor) score at 12 months between the DHA and placebo group using DASII. We performed the subgroup analysis using a two-sample *t*-test for the difference in mean DQ score between DHA and placebo group stratified by pre-specified subgroups, that is, mother's age (18–20, 21–25, 26–30, 31–35 years), mother's BMI (<18, 18.0–22.9, 23.0–24.9, ≥25 kg/m^2), gravidity (multigravida, primigravida), gestational age at delivery (<37, ≥37 weeks), duration of supplementation (≤10, 10.1–12.0, ≥12.1 months), vegetarian diet (yes, no), physical activity (inactive/low, moderate, high), and gender of the child (male, female). The *p*-value for interaction was calculated by including the interaction term between the characteristic of interest and treatment group in the linear regression model. Heterogeneity was assessed based on the significance of the interaction term between the characteristic of interest and treatment ($p < 0.05$). All the statistical analysis was done using STATA 16.0 version (College Station, TX, USA) and R 3.6.2 version.

3. Results

Enrollment began in January 2016 and ended in August 2017. Figure 1 shows the CONSORT flow chart. We screened 3379 women for eligibility and found 1131 to be eligible. A total of 957 mothers were enrolled and randomized (DHA = 478; Placebo = 479). Of the 902 live births, 878 were followed to 12 months and the DASII was administered to 863 of those. Loss to follow-up (8.2% from randomization through 12 months) and mean ± SD compliance did not differ by group (DHA: 93.0 ± 10.1; Placebo: 92.7 ± 10.7). Baseline maternal and household characteristics were similar by treatment group (Table 1). Overall, the mean age of the mothers was 23.9 (3.6) years, the mean BMI of the participating women was 20.6 ± 3.6 kg/m^2, and the mean hemoglobin was 11.1 ± 1.3 gm/L. Based on the semiquantitative FFQ administered at baseline, the dietary intake of n-3-rich foods was similar across the two groups. Additionally, there was no significant difference in the baseline characteristics between those who completed the study and those who did not complete the study (Table S1).

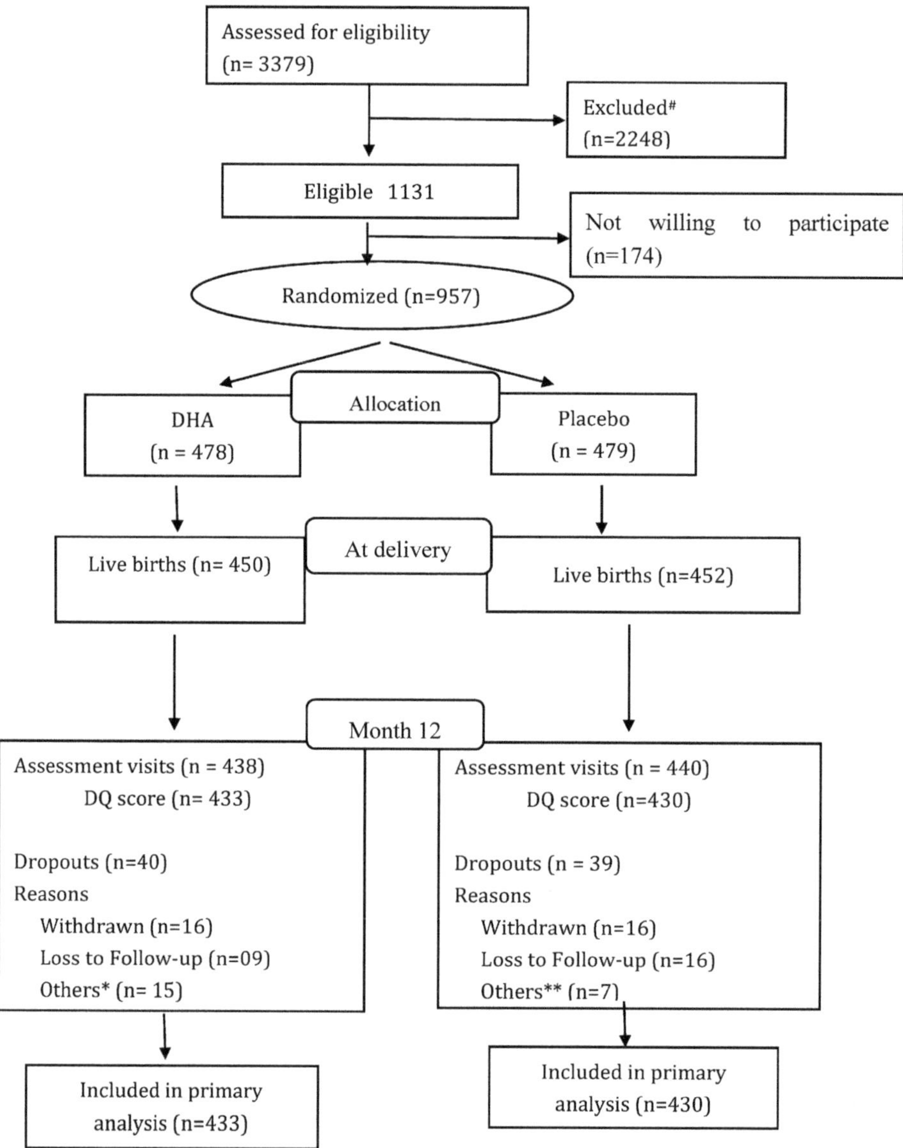

Figure 1. Consort. # Reasons for exclusion: gestational diabetes (n = 69); Hb < 7 gm% (n = 46); gestational age >20 weeks (n = 673); high-risk pregnancies (n = 118); chronic conditions (n = 246); under any other trial (n = 4); delivery plan other than PK (n = 835); missing/wrong contact information (n = 257). * Others included: abortion (n = 1); abruptio placenta (n = 1); fresh stillbirth (n = 4); macerated stillbirth (n = 3); neonatal death (n = 2); maternal death (n = 1); congenital anomalies (n = 1); infant death (n = 2) in DHA group. ** Others included: fresh stillbirth (n = 4); macerated stillbirth (n = 2); medical termination (n = 1) in Placebo group.

Table 1. Baseline Characteristics.

	DHA (*N* = 478)	**Placebo (*N* = 479)**
Maternal age (years), mean ± SD	23.5 ± 3.5	23.6 ± 3.7
Gestational age at enrollment (weeks), median (p25, p75)	15.0 (12, 18)	15.0 (12, 18)
Primigravida, n (%)	180 (37.7%)	206 (43.0%)
Education, n (%)		
College graduated and above	88 (18.4%)	82 (17.1%)
High school/secondary	371 (77.6%)	386 (80.6%)
Employed, n (%)	119 (25.0%)	104 (22.0%)
Household income (>Rs 20,000), n (%)	65 (13.6%)	47 (9.8%)
Dietary habits—vegetarian, n (%)	73 (15.3%)	87 (18.2%)
Consuming fish/seafood, n (%)	258 (53.9%)	202 (57.8%)
Anthropometric measurements		
Height (cm), mean ± SD	154.1 ± 5.6	153.9 ± 5.7
Weight (kg), mean ± SD	48.9 ± 9.0	48.9 ± 8.5
BMI (kg/m^2), mean ± SD	20.5 ± 3.5	20.7 ± 3.6
MUAC, (cm), mean ± SD	24.3 ± 3.0	24.3 ± 3.1
Biochemical measures		
Hb (gm%), mean ± SD	11.1 ± 1.3	11.2 ± 1.3
DHA (mol % of fatty acid) *-, mean ± SD	0.86 ± 0.78	0.88 ± 0.71

BMI: Body mass index; MUAC: Mid upper arm circumference; Hb: Hemoglobin; DHA: Docosahexaenoic acid; *- *N* = 258 (DHA); *N* = 224 (Placebo). Data are presented as mean ± standard deviation or median (p25, p75) or number (%).

The proportion of live births was similar across both the groups (DHA vs. Placebo: 94.1% vs. 94.3%) (Table 2). Initiation of breastfeeding in 1 h did not differ across groups (DHA vs. Placebo: 79.9% vs. 81.7%) (Table 3).

Table 2. Offspring characteristics at delivery.

At Delivery	**DHA** ***N* = 450**	**Placebo** ***N* = 452**
Live births, n (%)	450 (94.1%)	452 (94.3%)
Gestational age at the time of delivery (weeks), median (p25, p 75)	39.0 (38.0, 40.0)	39.0 (38.0, 40.0)
Delivery place—study center, n (%)	372 (84.7%)	375 (85.2%)
Delivery conducted by doctor, n (%)	416 (94.5%)	423 (96.1%)
Spontaneous labor, n (%)	410 (93.2%)	407 (92.5%)
Caesarean, n (%)	154 (35.0%)	175 (39.8%)
Male child, n (%)	234 (52.0%)	243 (54.0%)

Data are presented as median (p25, p75) or number (%); denominator for live births DHA (*N* = 478); Placebo (*N* = 479).

Table 3. Feeding practices.

	DHA n/N (%)	**Placebo** ***n/N* (%)**
Initiation of breastfeeding in 1 h, n (%)	346/433 (79.9%)	357/437 (81.7%)
Exclusively breastfed until 6 months, n (%)	236/426 (55.4%)	237/421 (56.3%)
Age at which complementary feeding initiated (months) *, mean ± SD	5.4 ± 0.67	5.3 ± 0.83

Data are presented as number (%); * DHA (*N* = 166); Placebo (*N* = 156).

The age at which complementary feeding started was similar across groups (DHA vs. Placebo, mean ± SD: 5.4 ± 0.67 vs. 5.3 ± 0.83 months) (Table 3).

The maternal RBC DHA concentrations (mol % of fatty acid) (mean ± SD) were not different between the groups at baseline (DHA; 0.86 ± 0.78, Placebo: 0.88 ± 0.71) but significantly higher in the DHA group compared to placebo group at delivery (DHA: 1.94 ± 1.42, Placebo: 0.84 ± 0.56; $p < 0.001$) (Table 4).

Table 4. Mean DHA (mol % of fatty acid) levels in RBC phospholipids.

	DHA	Placebo	Mean Difference * [95% CI]	*p*-Value
DHA at baseline	N = 258 0.86 ± 0.78, 0.56 (0.32, 1.21)	N = 224 0.88 ± 0.71, 0.55 (0.37, 1.28)	0.02 [0.11,0.15]	0.77
DHA at the time of delivery	N = 271 2.03 ± 1.75, 1.41 (0.61, 2.99)	N = 242 1.12 ± 0.86, 0.83 (0.42, 1.72)	−0.91 [−1.16, 0.67]	<0.001
DHA in cord blood	N = 265 2.61 ± 1.45, 2.64 (1.38, 3.81)	N = 232 1.83 ± 0.90, 1.76 (1.16, 2.48)	−0.77 [−0.99, 0.56]	<0.001
DHA in infant blood at 6 months	N = 263 1.94 ± 1.42, 1.65 (0.64, 3.08)	N = 240 0.84 ± 0.56, 0.77 (0.43, 1.12)	−1.09 [−1.29, −0.9]	<0.001
DHA in infant blood at 12 months	N = 227 1.71 ± 1.18, 1.50 (0.60, 2.64)	N = 204 1.27 ± 0.93, 1.03 (0.43, 1.89)	−0.44 [−0.64, −0.24]	<0.001

Data are presented as mean ± standard deviation, median (p25, p75); DHA: Docosahexaenoic acid; *p*-value calculated using unpaired *t*-test; * difference = Placebo minus DHA.

In the placebo group, there was no change in RBC DHA concentrations over the period of intervention while in the DHA group, concentrations increased by 1.04 (0.86, 1.23) mol % of fatty acid, paired *t*-test, $p < 0.001$.

The DQ score (mean ± SD) at 12 months for the DHA and placebo groups was 96.6 ± 12.2 and 97.1 ± 13.0, respectively, and the mean difference (placebo minus DHA) was 0.46 (95% CI: −1.23, 2.14; $p = 0.60$). The motor and mental scores also did not differ significantly across the groups at 12 months (Table 5). There were 25 serious adverse events, all determined to be unrelated to intervention (DHA group = 16; placebo = 9; $p = 0.154$) (Table S2). The last 12-month assessment (including DASII) was conducted in early April 2019.

Table 5. DQ score and components of DQ (motor and mental scores) at 12th month.

At 12th Month	DHA (n = 433)	Placebo (N = 430)	Difference * [95% CI]	*p*-Value
DQ score	96.6 ± 12.1	97.1 ± 13.0	0.46 [−1.23,2.14]	0.60
Motor score	47.6 ± 3.7	47.6 ± 3.7	0.03 [−0.47,0.52]	0.92
Mental score	106.0 ± 7.0	106.7 ± 7.6	0.63 [−0.35,1.61]	0.21

Data are presented as mean ± standard deviation and [95% CI]; unadjusted model: *t*-test for difference between mean values at 12 months; DQ: development quotient. * Placebo minus DHA.

The per protocol analysis also showed no difference in 12-month DQ scores between the two groups (Table S3). The mean infant DQ, mental and motor scores at six months did not differ significantly across groups (secondary outcomes not shown in this paper). The mean DQ score did not differ significantly across mother's age group, BMI, gravida, gestation age at delivery, compliance rate, duration of supplementation, physical activity (Figure 2).

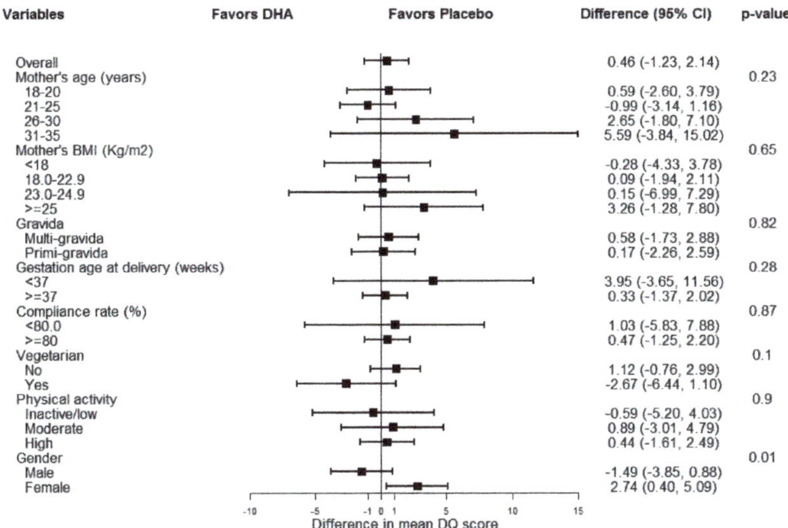

Figure 2. Subgroup analysis. Difference: Placebo minus DHA; Difference in mean DQ score between DHA and placebo group at 12 months was calculated using two-sample *t*-test for each subgroup; *p*-value for interaction calculated using linear regression model including interaction term for characteristic of interest and treatment group.

However, the mean DQ score for female child was higher in placebo as compared to DHA [mean difference (placebo– DHA) (95% CI): 2.74 (0.40, 5.09)]. The p-value for interaction was significant for gender of the child (*p* = 0.013). The pre- and postnatal maternal supplementation with 400 mg/d DHA did not improve the infant's DQ score as measured by DASII at 12 months of age.

4. Discussion

In this well designed and executed RCT (CONSORT checklist enclosed as Table S4), pre- and postnatal maternal supplementation with 400 mg/d DHA did not improve the infant's development score as measured by DASII at 12 months of age. Our findings are in accordance with some other trials [46–49]. The reviews published so far in this field to understand the association between maternal DHA supplementation and infant neurodevelopment have also reported the evidence of a positive association between maternal DHA supplementation and infant neurodevelopment to be either too low or inconsistent, with a majority of the RCTs showing no positive effect [36]. Makrides et al. provided 800 mg/d and [50] showed no difference in children's cognitive scores between the intervention and control group at 18 months. Similarly, Helland et al. [49] found no effect of prenatal supplementation with cod liver oil on cognitive development among 288 three-month-old children in Norway. On the other hand, the randomized trial of Colombo et al. [51] found that lower doses of DHA supplementation of the infants (4 to 9 months of age) showed better cognitive outcomes in terms of their attention span. Studies have reported maternal DHA status in pregnancy to be positively associated with infant's brain volume at 1 month [52], improved infant's attention [53], and enhanced problem-solving skills at 12 months [54]. Rees et al. [26] showed that the infants of mothers who had a higher dietary DHA intake during the second and third trimesters of their pregnancy performed better on cognitive assessment measures (habituation and sustained visual attention) at 4.5 and 9 months of age. However, it was also observed that the infants from the medium dietary DHA group [0.64% of fatty acids from DHA (34 mg/100 kcal)] performed significantly better than the low-DHA [0.32% (17 mg/100 kcal)] and high-DHA group [0.96% (51 mg/100 kcal)], with the latter showing the worst performance [26,51].

Additionally, some studies have reported improved cognition among older children whose mothers were supplemented with 400–600 mg/d DHA prenatally, after they turned four years old [27,30,35].

The neurodevelopment in the current study was assessed with the help of the DASII test which is BSID's adaptation. Although BSID is a global standardized test for assessing cognitive functioning and has been frequently utilized in neurodevelopment effects of LCPUFA in infants [55], its sensitivity to pick up some subtle differences in infant cognitive ability has been questioned [56]. Experts suggest that differences in intellectual functioning that are sensitive to pathways influenced by n-3 fatty acids may only be detected with more sensitive measures of neurodevelopment such as neuroimaging techniques [57], neuropsychological testing [58], distinct cognitive abilities, or executive functioning [59]. The suitability of these approaches for large field-based trials in low resource settings warrants further exploration.

Few possible limitations of our single-center trial may include lack of details on home environment and detailed dietary data on the children which may have influenced the neurodevelopment in the first year of life. The measure of neurodevelopment used in the study (i.e., the DQ score) has not been validated against the functional magnetic resonance imaging (fMRI) which is currently considered the gold standard noninvasive hemodynamic-based neuroimaging technology. The fatty acid composition of the blood and breast milk collected from the enrolled women are not currently available. We also did not estimate dietary n-3 LCPUFA intakes in all the study subjects; however, the biomarker (RBC phospholipid DHA) is very responsive to intake and was measured at both enrollment and immediately after birth as an indicator of prior status and study DHA intake. Further, other fatty acids like AA and DHA:AA ratios may play a key role in neurodevelopment [60]. To the best of our knowledge, our study is the first one to examine the effects of in utero and early-life DHA exposure (through maternal supplementation from mid-pregnancy through six months postpartum) on postnatal neurodevelopment of Indian infants. Through a long supplementation phase and follow-up period, the participants of our trial reported good compliance and very low attrition. Genetic predisposition, prenatal and postnatal care and nutrition, and social and physical environment may all be critical in shaping the neurodevelopment of an infant. We collected maternal dietary data at baseline but not at later visits. Changes throughout the pregnancy with respect to not only n-3 LCPUFA but also other nutrients such as vitamin D or iron [61,62] may affect offspring neurodevelopment.

While DHA is a key intrinsic factor constituting more than 40% of brain polyunsaturated fatty acids [63,64], external factors like care and stimulation in the home environment also play a significant role in a child's cognitive development [65,66]. Poor physical conditions of home and limited access to age-appropriate learning materials have been linked with social–emotional problems in children. Ramakrishnan et al. [46] indicated a possible attenuating effect of DHA on the positive association between home environment and psychomotor development. DHA may be especially helpful for children living in home environments characterized by reduced caregiver interactions and opportunities for early childhood stimulation [67]. It might have been useful to have these data in the present study. Further, the mean DQ score of girl children in the placebo group was higher. This inconsistency may be due to chance. Some studies also attribute gender-related differences in neurodevelopment to sex hormones [68] and/or social context [69], but since these were not assessed in our trial, we cannot be certain of this.

In summary, supplementing mothers through pregnancy and lactation with 400 mg/d DHA (vs. placebo) did not benefit offspring neurodevelopment at one year of age in this Indian setting. Deeper insights into maternal dietary patterns, young child feeding practices, home environment, and the interactions amongst these factors are warranted to understand what shapes early neurodevelopment.

Supplementary Materials: The following are available online at http://www.mdpi.com/2072-6643/12/10/3041/s1, Table S1: Comparison of baseline characteristics of those who completed the study and those who did not, Table S2: Listing of serious adverse events by group, Table S3: Primary outcome (per protocol analysis), Table S4: CONSORT checklist.

Author Contributions: S.K. had full access to all the data in the study and takes responsibility for the integrity of the data and the accuracy of the data analysis. She also takes the final responsibility for the decision to submit for publication. Conceptualization, S.K., D.P., N.T., U.R., and A.D.S.; Data curation, D.K.; Formal analysis, D.K. and R.G.; Funding acquisition, S.K. and D.P.; Investigation, S.K., D.M., and S.J.; Methodology, U.R.; Project administration, S.K., M.C., and K.P.; Resources, K.P., M.K.S., and D.P.; Software, D.K.; Supervision, K.P., M.K.S., M.K., N.T., and A.D.S.; Validation, R.G.; Writing—original draft, S.K.; Writing—review & editing, M.C., G.D., D.P., N.T., U.R., and A.D.S. All authors have read and agreed to the published version of the manuscript.

Funding: This trial is funded by DBT/Wellcome Trust India Alliance (India Alliance) (December 2015–December 2020) via Grant number IA/CPHE/14/1/501498. The Young Scientist Award by DST SERB India (2013–16) via grant number SR/FT/LS-156/2011 helped us establish a part of the DHANI trial cohort. The funding sources had no role in the design and conduct of the study; collection, management, analysis, and interpretation of the data; preparation, review, or approval of the manuscript; and decision to submit the manuscript for publication. Trial Registration—ClinicalTrials.gov Identifier NCT03072277; Ctri.nic.in Identifier CTRI/2017/08/009296.

Acknowledgments: The trial's mentoring advisory committee (MAC) chairs Professors Reynaldo Martorell (Emory University, Atlanta) and K Srinath Reddy (PHFI, India) have provided overall guidance at each stage and we are extremely grateful for that. All other members of the MAC also made significant contributions in the conduct of the study. The Department of Obstetrics and Gynecology at KAHER's JN Medical College (JNMC) Belagavi is deeply acknowledged. The Unit Chiefs (M. B. Bellad, Anita Dalal, Yeshita Pujar, and M. C. Metgud) deserve a special mention in providing valuable inputs and support during each phase of this study. We are indebted to the JNMC's Blood Bank in-charge Mr. Shrikant Viragi and his whole team for facilitating all onsite biochemical work in the study. The efforts and diligence of IRBs of all participating institutions are also acknowledged. We express sincere gratitude to the Data Safety and Monitoring Board (DSMB) chaired by Prof Vinod Paul (Niti Aayog, India). We are truly grateful to the resource and support teams at Centre for Chronic Disease Control (CCDC) and Public Health Foundation of India (PHFI), Delhi. The active ingredient DHA-S (also known as 'DHA algal oil') is a naturally occurring, microalgal oil derived from Schizochytrium sp. Each active 635 mg softgel provided ~200 mg DHA (total daily DHA = 400 mg/d) and a 50/50 corn oil and soybean oil blend was used for the placebo softgels (DSM Nutritional Products, Columbia, MD, USA. Neither the product provider nor the sponsors had an opportunity to review a presubmission copy of the article.

Conflicts of Interest: The authors declare no conflict of interest.

Data Sharing Statement: The request for accessing de-identified data plus data dictionary will be put forth for approval by the trial's mentoring and advisory committee (MAC). Interested researchers may submit a proposal with a valid reason or justification (e.g., meta-analysis, etc.).

References

1. Grantham-McGregor, S.; Cheung, Y.B.; Cueto, S.; Glewwe, P.; Richter, L.; Strupp, B. International Child Development Steering Group. Developmental potential in the first 5 years for children in developing countries. *Lancet* **2007**, *369*, 60–70. [CrossRef] [PubMed]

2. Jiang, X.; Nardelli, J. Cellular and molecular introduction to brain development. *Neurobiol. Dis.* **2016**, *92 Pt A*, 3–17. [CrossRef]

3. Cusick, S.; Georgieff, M.K. The Role of Nutrition in Brain Development: The Golden Opportunity of the "First 1000 Days". *J. Pediatr.* **2016**, *175*, 16–21. [CrossRef] [PubMed]

4. Costello, A.M.D.L.; Osrin, D. Micronutrient Status during Pregnancy and Outcomes for Newborn Infants in Developing Countries. *J. Nutr.* **2003**, *133* (Suppl. 2), 1757S–1764S. [CrossRef] [PubMed]

5. Larqué, E.; Demmelmair, H.; Koletzko, B. Perinatal supply and metabolism of long-chain polyunsaturated fatty acids: Importance for the early development of the nervous system. *Ann. New York Acad. Sci.* **2002**, *967*, 299–310.

6. Rahmanifar, A.; Kirksey, A.; Wachs, T.D.; McCabe, G.P.; Bishry, Z.; Galal, O.M.; Harrison, G.G.; Jerome, N.W. Diet during lactation associated with infant behavior and caregiver-infant interaction in a semirural Egyptian village. *J. Nutr.* **1993**, *123*, 164–175.

7. Meldrum, S.; Karen, S. Docosahexaenoic Acid and Neurodevelopmental Outcomes of Term Infants. *Ann. Nutr. Metab.* **2016**, *69* (Suppl. 1), 22–28. [CrossRef] [PubMed]

8. Makrides, M. DHA supplementation during the perinatal period and neurodevelopment: Do some babies benefit more than others? *Prostaglandins Leukot. Essent. Fat. Acids* **2013**, *88*, 87–90. [CrossRef]

9. Ryan, A.S.; Astwood, J.D.; Gautier, S.; Kuratko, C.N.; Nelson, E.B.; Salem, N. Effects of long-chain polyunsaturated fatty acid supplementation on neurodevelopment in childhood: A review of human studies. *Prostaglandins Leukot. Essent. Fat. Acids* **2010**, *82*, 305–314. [CrossRef]

10. Makrides, M.; Smithers, L.G.; Gibson, R. Role of Long-Chain Polyunsaturated Fatty Acids in Neurodevelopment and Growth. *Nestle Nutr. Workshop Ser. Pediatr Program* **2010**, *65*, 123–136.

11. Hibbeln, J.R.; Davis, J.M.; Steer, C.; Emmett, P.; Rogers, I.; Williams, C.; Golding, J. Maternal seafood consumption in pregnancy and neurodevelopmental outcomes in childhood (ALSPAC study): An observational cohort study. *Lancet* **2007**, *369*, 578–585. [CrossRef]

12. Docosahexaenoic acid (DHA). Monograph. *Altern. Med. Rev.* **2009**, *14*, 391–399.

13. Lauritzen, L.; Brambilla, P.; Mazzocchi, A.; Harsløf, L.B.S.; Ciappolino, V.; Agostoni, C. DHA Effects in Brain Development and Function. *Nutrents* **2016**, *8*, 6. [CrossRef]

14. Carlson, S.E. Docosahexaenoic acid supplementation in pregnancy and lactation. *Am. J. Clin. Nutr.* **2009**, *89*, 678s–684s. [CrossRef]

15. Makrides, M.; Gibson, R. Long-chain polyunsaturated fatty acid requirements during pregnancy and lactation. *Am. J. Clin. Nutr.* **2000**, *71* (Suppl. 1), 307S–311S. [CrossRef]

16. Koletzko, B.; Boey, C.C.; Campoy, C.; Carlson, S.E.; Chang, N.; Guillermo-Tuazon, M.A.; Joshi, S.; Prell, C.; Quak, S.H.; Sjarif, D.R.; et al. Current Information and Asian Perspectives on Long-Chain Polyunsaturated Fatty Acids in Pregnancy, Lactation, and Infancy: Systematic Review and Practice Recommendations from an Early Nutrition Academy Workshop. *Ann. Nutr. Metab.* **2014**, *65*, 49–80. [CrossRef]

17. Weiser, M.J.; Butt, C.M.; Mohajeri, M.H. Docosahexaenoic Acid and Cognition throughout the Lifespan. *Nutrients* **2016**, *8*, 99. [CrossRef]

18. Garg, P.; Pejaver, R.K.; Sukhija, M.; Ahuja, A. Role of DHA, ARA, & phospholipids in brain development: An Indian perspective. *Clin. Epidemiology Glob. Health* **2017**, *5*, 155–162. [CrossRef]

19. Brenna, J.T. Animal studies of the functional consequences of suboptimal polyunsaturated fatty acid status during pregnancy, lactation and early post-natal life. *Matern. Child Nutr.* **2011**, *7* (Suppl. 2), 59–79. [CrossRef]

20. Wadhwani, N.; Patil, V.; Joshi, S.R. Maternal long chain polyunsaturated fatty acid status and pregnancy complications. *Prostaglandins Leukot. Essent. Fat. Acids* **2018**, *136*, 143–152. [CrossRef]

21. Wierzejska, R.; Jarosz, M.; Wojda, B.; Siuba-Strzelińska, M. Dietary intake of DHA during pregnancy: A significant gap between the actual intake and current nutritional recommendations. *Roczniki Państwowego Zakładu Higieny* **2018**, *69*, 381–386. [CrossRef]

22. Kris-Etherton, P.M.; Grieger, J.A.; Etherton, T.D. Dietary reference intakes for DHA and EPA. *Prostaglandins Leukot. Essent. Fat. Acids* **2009**, *81*, 99–104. [CrossRef]

23. Huffman, S.L.; Harika, R.K.; Eilander, A.; Osendarp, S.J. Essential fats: How do they affect growth and development of infants and young children in developing countries? A literature review. *Matern. Child Nutr.* **2011**, *7* (Suppl. 3), 44–65. [CrossRef]

24. Shrimpton, R.; Huffman, S.L.; Zehner, E.R.; Darnton-Hill, I.; Dalmiya, N. Multiple Micronutrient Supplementation during Pregnancy in Developing-Country Settings: Policy and Program Implications of the Results of a Meta-Analysis. *Food Nutr. Bull.* **2009**, *30* (Suppl. 4), S556–S573. [CrossRef]

25. Muthayya, S.; Dwarkanath, P.; Thomas, T.; Ramprakash, S.; Mehra, R.; Mhaskar, A.; Mhaskar, R.; Thomas, A.; Bhat, S.; Vaz, M.; et al. The effect of fish and ω-3 LCPUFA intake on low birth weight in Indian pregnant women. *Eur. J. Clin. Nutr.* **2009**, *63*, 340–346. [CrossRef]

26. Rees, A.; Sirois, S.; Wearden, A. Prenatal maternal docosahexaenoic acid intake and infant information processing at 4.5mo and 9mo: A longitudinal study. *PLoS ONE* **2019**, *14*, e0210984. [CrossRef]

27. Colombo, J.; Shaddy, D.J.; Gustafson, K.M.; Gajewski, B.J.; Thodosoff, J.M.; Kerling, E.H.; Carlson, S.E. The Kansas University DHA Outcomes Study (KUDOS) clinical trial: Long-term behavioral follow-up of the effects of prenatal DHA supplementation. *Am. J. Clin. Nutr.* **2019**, *109*, 1380–1392. [CrossRef]

28. Argaw, A.; Huybregts, L.; Wondafrash, M.; Kolsteren, P.; Belachew, T.; Worku, B.N.; Abessa, T.G.; Bouckaert, K.P. Neither n-3 Long-Chain PUFA Supplementation of Mothers through Lactation nor of Offspring in a Complementary Food Affects Child. Overall or Social-Emotional Development: A 2 × 2 Factorial Randomized Controlled Trial in Rural Ethiopia. *J. Nutr.* **2019**, *149*, 505–512. [CrossRef]

29. Gould, J.F.; Yelland, L.N.; Smithers, L.G.; Makrides, M.; Treyvaud, K.; Anderson, V.; McPhee, A. Seven-Year Follow-up of Children Born to Women in a Randomized Trial of Prenatal DHA Supplementation. *JAMA* **2017**, *317*, 1173–1175. [CrossRef]

30. Ramakrishnan, U.; Gonzalez-Casanova, I.; Schnaas, L.; DiGirolamo, A.; Quezada, A.D.; Pallo, B.C.; Hao, W.; Neufeld, L.M.; Rivera, J.A.; Stein, A.D.; et al. Prenatal supplementation with DHA improves attention at 5 y of age: A randomized controlled trial. *Am. J. Clin. Nutr.* **2016**, *104*, 1075–1082. [CrossRef]

31. Makrides, M.; Gould, J.F.; Gawlik, N.R.; Yelland, L.N.; Smithers, L.G.; Anderson, P.J.; Gibson, R.A. Four-Year Follow-up of Children Born to Women in a Randomized Trial of Prenatal DHA Supplementation. *JAMA* **2014**, *311*, 1802–1804. [CrossRef]

32. Carlson, S.E.; Colombo, J.; Gajewski, B.J.; Gustafson, K.M.; Mundy, D.; Yeast, J.; Georgieff, M.K.; Markley, L.A.; Kerling, E.H.; Shaddy, D.J. DHA supplementation and pregnancy outcomes. *Am. J. Clin. Nutr.* **2013**, *97*, 808–815. [CrossRef]

33. Escolano-Margarit, M.V.; Ramos, R.; Beyer, J.; Csábi, G.; Parrilla-Roure, M.; Cruz, F.; Pérez-García, M.; Hadders-Algra, M.; Gil, Á; Decsi, T.; et al. Prenatal DHA Status and Neurological Outcome in Children at Age 5.5 Years Are Positively Associated. *J. Nutr.* **2011**, *141*, 1216–1223. [CrossRef]

34. Ramakrishnan, U.; Stein, A.D.; Parra-Cabrera, S.; Wang, M.; Imhoff-Kunsch, B.; Juárez-Márquez, S.; Rivera, J.; Martorell, R. Effects of Docosahexaenoic Acid Supplementation During Pregnancy on Gestational Age and Size at Birth: Randomized, Double-Blind, Placebo-Controlled Trial in Mexico. *Food Nutr. Bull.* **2010**, *31*, S108–S116. [CrossRef]

35. Gustafson, K.M.; Liao, K.; Mathis, N.B.; Shaddy, D.J.; Kerling, E.H.; Christifano, D.N.; Colombo, J.; Carlson, S.E. Prenatal docosahexaenoic acid supplementation has long-term effects on childhood behavioral and brain responses during performance on an inhibitory task. *Nutr. Neurosci.* **2020**, 1–11. [CrossRef]

36. Newberry, S.J.; Chung, M.; Booth, M.; Maglione, M.A.; Tang, A.M.; O'Hanlon, C.E.; Wang, D.D.; Okunogbe, A.; Huang, C.; Motala, A.; et al. Omega-3 Fatty Acids and Maternal and Child Health: An Updated Systematic Review. *Évid. Rep. Assess.* **2016**, *2016*, 1–826. [CrossRef]

37. Bergmann, R.L.; Haschke-Becher, E.; Klassen-Wigger, P.; Bergmann, K.E.; Richter, R.; Dudenhausen, J.W.; Grathwohl, D.; Haschke, F. Supplementation with 200 mg/day docosahexaenoic acid from mid-pregnancy through lactation improves the docosahexaenoic acid status of mothers with a habitually low fish intake and of their infants. *Ann. Nutr. Metab.* **2008**, *52*, 157–166. [CrossRef]

38. Khandelwal, S.; Swamy, M.K.; Patil, K.; Kondal, D.; Chaudhry, M.; Gupta, R.; Divan, G.; Kamate, M.; Ramakrishnan, L.; Bellad, M.B.; et al. The impact of DocosaHexaenoic Acid supplementation during pregnancy and lactation on Neurodevelopment of the offspring in India (DHANI): Trial protocol. *BMC Pediatr.* **2018**, *18*, 261. [CrossRef]

39. Longvah, T.; Bhaskarachary, K.; Venkaiah, K.; Longvah, T.; Anantan, I. *Indian food Composition Tables*; National Institute of Nutrition: Telangana, India, 2017.

40. Patni, B. Developmental Assessment Scales for Indian Infants (DASII). *Ind. J. Prac. Pediatr.* **2012**, *14*, 409–412.

41. Ferreri, C.; Masi, A.; Sansone, A.; Giacometti, G.; LaRocca, A.V.; Menounou, G.; Scanferlato, R.; Tortorella, S.; Rota, D.; Conti, M.; et al. Fatty Acids in Membranes as Homeostatic, Metabolic and Nutritional Biomarkers: Recent Advancements in Analytics and Diagnostics. *Diagnostics* **2016**, *7*, 1. [CrossRef]

42. Matthan, N.R.; Ooi, E.M.; Van Horn, L.; Neuhouser, M.L.; Woodman, R.; Lichtenstein, A.H. Plasma Phospholipid Fatty Acid Biomarkers of Dietary Fat Quality and Endogenous Metabolism Predict Coronary Heart Disease Risk: A Nested Case-Control Study Within the Women's Health Initiative Observational Study. *J. Am. Hear. Assoc.* **2014**, *3*, e000764. [CrossRef]

43. Rose, H.G.; Oklander, M. Improved Procedure for The Extraction of Lipids from Human Erythrocytes. *J. Lipid Res.* **1965**, *6*, 428–431.

44. Lepage, G.; Roy, C.C. Direct transesterification of all classes of lipids in a one-step reaction. *J. Lipid Res.* **1986**, *27*, 114–120.

45. Finley, D.A.; Lönnerdal, B.; Dewey, K.G.; E Grivetti, L. Breast milk composition: Fat content and fatty acid composition in vegetarians and non-vegetarians. *Am. J. Clin. Nutr.* **1985**, *41*, 787–800. [CrossRef]

46. Ramakrishnan, U.; Stinger, A.; DiGirolamo, A.M.; Martorell, R.; Neufeld, L.M.; Rivera, J.A.; Schnaas, L.; Stein, A.D.; Wang, M. Prenatal Docosahexaenoic Acid Supplementation and Offspring Development at 18 Months: Randomized Controlled Trial. *PLoS ONE* **2015**, *10*, e0120065. [CrossRef]

47. Van Der Merwe, L.F.; Moore, S.E.; Fulford, A.J.; Halliday, K.E.; Drammeh, S.; Young, S.; Prentice, A.M. Long-chain PUFA supplementation in rural African infants: A randomized controlled trial of effects on gut integrity, growth, and cognitive development. *Am. J. Clin. Nutr.* **2013**, *97*, 45–57. [CrossRef]

48. Van Goor, S.A.; Dijck-Brouwer, D.J.; Erwich, J.J.H.; Schaafsma, A.; Hadders-Algra, M. The influence of supplemental docosahexaenoic and arachidonic acids during pregnancy and lactation on neurodevelopment at eighteen months. *Prostaglandins Leukot. Essent. Fat. Acids* **2011**, *84*, 139–146. [CrossRef]

49. Helland, I.B.; Smith, L.; Sareem, K.; Saugstad, O.D.; Drevon, C.A. Maternal supplementation with very-long-chain n-3 fatty acids during pregnancy and lactation augments children's IQ at 4 years of age. *Pediatrics* **2003**, *111*, e39–e44.

50. Makrides, M.; Gibson, R.; McPhee, A.J.; Yelland, L.; Quinlivan, J.; Ryan, P. Effect of DHA Supplementation During Pregnancy on Maternal Depression and Neurodevelopment of Young Children. *JAMA* **2010**, *304*, 1675–1683. [CrossRef]

51. Colombo, J.; Carlson, S.E.; Cheatham, C.L.; Gustafson, K.M.; Kepler, A.; Doty, T. Long-chain polyunsaturated fatty acid supplementation in infancy reduces heart rate and positively affects distribution of attention. *Pediatr. Res.* **2011**, *70*, 406–410. [CrossRef]

52. Morton, S.U.; Vyas, R.; Gagoski, B.; Vu, C.; Litt, J.; Larsen, R.J.; Kuchan, M.J.; Lasekan, J.B.; Sutton, B.P.; Grant, P.E.; et al. Maternal Dietary Intake of Omega-3 Fatty Acids Correlates Positively with Regional Brain Volumes in 1-Month-Old Term Infants. *Cereb. Cortex* **2020**, *30*, 2057–2069. [CrossRef] [PubMed]

53. Colombo, J.; Gustafson, K.M.; Gajewski, B.J.; Shaddy, D.J.; Kerling, E.H.; Thodosoff, J.M.; Doty, T.; Brez, C.C.; Carlson, S.E. Prenatal DHA supplementation and infant attention. *Pediatr. Res.* **2016**, *80*, 656–662. [CrossRef] [PubMed]

54. Braarud, H.C.; Markhus, M.W.; Skotheim, S.; Stormark, K.M.; Frøyland, L.; Graff, I.E.; Kjellevold, M. Maternal DHA Status during Pregnancy Has a Positive Impact on Infant Problem Solving: A Norwegian Prospective Observation Study. *Nutrients* **2018**, *10*, 529. [CrossRef]

55. Gould, J.F.; Smithers, L.G.; Makrides, M. The effect of maternal omega-3 (n-3) LCPUFA supplementation during pregnancy on early childhood cognitive and visual development: A systematic review and meta-analysis of randomized controlled trials. *Am. J. Clin. Nutr.* **2013**, *97*, 531–544. [CrossRef] [PubMed]

56. Colombo, J.; Carlson, S.E. Is the measure the message: The BSID and nutritional interventions. *Pediatrics* **2012**, *129*, 1166–1167. [CrossRef] [PubMed]

57. Jensen, C.L.; Voigt, R.G.; Llorente, A.M.; Peters, S.U.; Prager, T.C.; Zou, Y.L.; Rozelle, J.C.; Turcich, M.R.; Fraley, J.K.; Anderson, R.E.; et al. Effects of Early Maternal Docosahexaenoic Acid Intake on Neuropsychological Status and Visual Acuity at Five Years of Age of Breast-Fed Term Infants. *J. Pediatr.* **2010**, *157*, 900–905. [CrossRef]

58. Bradshaw, J.L. *Developmental Disorders of the Frontostriatal System: Neuropsychological, Neuropsychiatric and Evolutionary Perspectives. Developmental Disorders of the Frontostriatal System: Neuropsychological, Neuropsychiatric and Evolutionary Perspectives*; Psychology Press: New York, NY, USA, 2001.

59. Cheatham, C.L.; Colombo, J.; Carlson, S.E. N-3 fatty acids and cognitive and visual acuity development: Methodologic and conceptual considerations. *Am. J. Clin. Nutr.* **2006**, *83*, 1458S–1466S. [CrossRef]

60. Klevebro, S.; Juul, S.E.; Wood, T.R. A More Comprehensive Approach to the Neuroprotective Potential of Long-Chain Polyunsaturated Fatty Acids in Preterm Infants Is Needed—Should We Consider Maternal Diet and the n-6:n-3 Fatty Acid Ratio? *Front. Pediatr.* **2020**, *7*. [CrossRef]

61. Santos, D.C.C.; Angulo-Barroso, R.; Li, M.; Bian, Y.; Sturza, J.; Richards, B.; Lozoff, B. Timing, duration, and severity of iron deficiency in early development and motor outcomes at 9 months. *Eur. J. Clin. Nutr.* **2018**, *72*, 332–341. [CrossRef]

62. Morse, N.L. Benefits of Docosahexaenoic Acid, Folic Acid, Vitamin D and Iodine on Foetal and Infant Brain Development and Function Following Maternal Supplementation during Pregnancy and Lactation. *Nutrents* **2012**, *4*, 799–840. [CrossRef]

63. Singh, M. Essential fatty acids, DHA and human brain. *Indian J. Pediatr.* **2005**, *72*, 239–242. [CrossRef] [PubMed]

64. Dyall, S.C. Long-chain omega-3 fatty acids and the brain: A review of the independent and shared effects of EPA, DPA and DHA. *Front. Aging Neurosci.* **2015**, *7*, 52. [CrossRef] [PubMed]

65. Ferguson, K.T.; Cassells, R.C.; MacAllister, J.W.; Evans, G.W. The physical environment and child development: An international review. *Int. J. Psychol.* **2013**, *48*, 437–468. [CrossRef] [PubMed]

66. Susane, A.A.; Santos, D.N.; Bastos, A.C.; Pedromonico, M.R.M.; de Almeida-Filho, N.; Barreto, M.L. Family environment and child's cognitive development: An epidemiological approach. *Rev. Saude Publica* **2005**, *39*, 1–6.

67. Nguyen, P.H.; DiGirolamo, A.M.; Young, M.; Kim, N.; Nguyen, S.; Martorell, R.; Ramakrishnan, U.; Gonzalez-Casanova, I. Influences of early child nutritional status and home learning environment on child development in Vietnam. *Matern. Child Nutr.* **2018**, *14*, e12468. [CrossRef]

68. Pinares-Garcia, P.; Stratikopoulos, M.; Zagato, A.; Loke, H.; Lee, J. Sex: A Significant Risk Factor for Neurodevelopmental and Neurodegenerative Disorders. *Brain Sci.* **2018**, *8*, 154. [CrossRef]

69. Lemmers-Jansen, I.L.J.; Krabbendam, L.; Veltman, D.J.; Fett, A.-K.J. Boys vs. girls: Gender differences in the neural development of trust and reciprocity depend on social context. *Dev. Cogn. Neurosci.* **2017**, *25*, 235–245. [CrossRef]

MDPI
St. Alban-Anlage 66
4052 Basel
Switzerland
Tel. +41 61 683 77 34
Fax +41 61 302 89 18
www.mdpi.com

Nutrients Editorial Office
E-mail: nutrients@mdpi.com
www.mdpi.com/journal/nutrients

www.ingramcontent.com/pod-product-compliance
Lightning Source LLC
LaVergne TN
LVHW070626170726
843515LV00004B/185